Catalytic Reforming

Catalytic Reforming

Donald M. Little

PennWell Publishing Company
Tulsa, Oklahoma

Copyright © 1985 by
PennWell Publishing Company
1421 South Sheridan Road/P.O. Box 1260
Tulsa, Oklahoma 74101

All rights reserved. No part of this book may be
reproduced, stored in a retrieval system, or
transcribed in any form or by any means, electronic
or mechanical, including photocopying and recording,
without the prior written permission of the publisher.

Printed in the United States of America

Library of Congress Cataloging in Publication Data

Little, Donald M.
 Catalytic reforming.

 Includes index.
 1. Catalytic reforming. I. Title.
TP690.4.L57 1985   665.5′3   84-26635
ISBN 0-87814-281-9

Table of Contents

Preface ix

Acknowledgments xi

Introduction xiii

Nomenclature and Units xvii

    Chapter 1 *Catalytic Reforming in Refinery Processing* *1*
        Products from Reforming 1
        Feedstocks for Reforming 3
        Octane Numbers 4
        Catalytic Reforming for Aromatics 12
        Summary 12

    Chapter 2 *How Catalytic Reformers Work—Chemical Reactions* *15*
        Reforming Reactions 15
        Feedstock Characterization 24
        Summary 26

    Chapter 3 *How Catalytic Reformers Work—Process Design* *28*
        Feed Preparation 28
        Temperature Control 31
        Reactors 31
        Product Recovery 32
        Reformer Classifications 33
        Reactor Design 35
        Recycle Compressor 38
        Summary 38

    Chapter 4 *The Catalyst* *40*
        Early Development 40
        Activity 41
        Selectivity 54
        Stability 55

Catalyst Development   57
Catalyst Preparation   58
Multifunction Catalyst   59
Catalyst Composition   60
Commercial Catalysts   61
Catalyst Poisons   68
Summary   70

Chapter 5   *Process Variables and Unit Operation*   *73*
Feedstock Properties   73
Reaction Temperature   79
Space Velocity   83
Reaction Pressure   87
Hydrogen-to-Hydrocarbon Ratio   90
Catalyst Type   91
Monitoring Reformer Operation   91
Chloride and Water Control   106
Troubleshooting   110
Loss of Octane   110
Loss of Yield   112
Loss of Delta T   113
Reformate Color   114
Fouling and Corrosion   116
Reforming Cracked Naphthas   117
Regeneration   128
Shutdown and Purge   129
Carbon Burn   129
Proof Burn   133
Rejuvenation of Catalyst   134
Reduction   134
Sulfiding and Start-up   134
Merchant Regeneration   136
Safety   137
Summary   139

Chapter 6   *BTX Operation*   *142*
Feedstock for BTX Reforming   143
Equilibrium Distribution of Aromatics   143
Process Variables   146
Laboratory Analysis for Aromatics   150
Summary   150

Chapter 7   *Commercial Processes*   *153*
IFP Catalytic Reforming   155
Magnaforming   158
Platforming   160
Powerforming   163
Rheniforming   166
Ultraforming   169
STAR   171
Summary   173

TABLE OF CONTENTS vii

Chapter 8 *Feed Preparation—Naphtha Hydrotreating 175*
Chemical Reactions 176
Hydrotreater Design 176
Operating Conditions 180
Hydrotreating Catalyst 181
Metals Adsorption on Catalyst 184
Sulfiding HDS Catalyst 185
Monitoring Naphtha Hydrotreating 186
Troubleshooting an HDS Unit 186
Recombination of Hydrogen Sulfide 187
Coking 188
Nitrogen 188
Silica 188
Deposits on Catalyst 189
Chloride Deposit and Corrosion 189
Safety 190
Catalyst Support Materials 191
Summary 192

Appendix I Octane Numbers of Selected Hydrocarbons 195
   II Physical Properties of Important Petrochemical Materials 198
   III Wire Mesh Openings (U.S. Sieve Series) 202
   IV Temperature Conversion Chart 203
   V 1984 Licensors of Commercial Catalytic Reforming Processes 208
   VI Manufacturers of Reforming Catalysts 209
   VII Hydrocarbons in Straight-run Base Stocks 210
   VIII Hydrocarbons in Midcontinent Naphthas 214
   IX Properties of Naphthas in World Crude Oils 215

Glossary 223

Index 231

Preface

Catalytic reforming is a key process in petroleum refining. It is the one process which makes high-octane-number unleaded gasoline a reality. This is the one process which converts, or reconstructs, without changing carbon numbers in the molecule, gasoline-boiling range low-octane hydrocarbons to the high-octane gasoline components required to fuel automobiles equipped with catalytic converters and emission controls. The push is on in 1985 to well nigh eliminate lead from gasoline, both in the United States and in Europe. Although oxygenated hydrocarbons, for example, alcohols and methyltertiarybutylether, have their uses, refiners depend on catalytic reforming for conversion of low-octane naphtha to high-octane gasoline. In the 1970s, revamp of reformers already in service and bimetallic catalysts enabled refiners to meet lead phasedown. The 1985 proposed reduction of lead to 0.1 gm/gal or less will give impetus to a surge of new low-pressure reformers designed to produce reformate of 100 RON clear or higher.

The high-octane-number hydrocarbons in the gasoline boiling range, produced in reforming, are primarily *aromatic* hydrocarbons, such as benzene, toluene, orthoxylene, metaxylene, paraxylene, and C_9^+ aromatics. The aromatics are not only premium blending stocks for motor fuel, but some are also in demand as petrochemicals, from which a wide range of plastic film and fibers are produced, as well as solvents and adhesives. In fact, catalytic reforming is the source of 85 to 90% of aromatics for chemical use in the United States.

There are a number of catalytic reforming units specifically designed and operated to produce *benzene, toluene,* and *xylenes* for the chemical market. Such reformers are generally called *BTX* units.

One other product of consequence from catalytic reforming is *hydrogen.* A reformer operating with average reactor pressure of 100 psig will produce from 1,000 to 1,500 standard cubic feet (scf) of hydrogen per barrel of charge, or from 20 to 30 million (MM) scf/day of hydrogen for a 20,000 barrels per stream day (b/sd) reformer. Catalytic reforming thus supplies large volumes of hydrogen—essential for desulfurization, denitrification, and olefin saturation in catalytic hydrotreating of naphtha, jet fuel, diesel fuel, gas oil, and residual oils.

Catalyst is the key to the catalytic reforming process. Reforming catalyst is a marvel in its design and ability to perform specifically desired chemical reactions and to essentially exclude undesirable reactions.

In a single-reaction system, the catalyst performs both dehydrogenation and hydrogenation, isomerization, hydrocracking, and dehydrocyclization reactions with incredible efficiency and stability. When the catalyst has been fouled with coke, it can be regenerated, rejuvenated, and restored to its original performance over and over again.

The catalysts are comprised of various combinations of crystallites of platinum, rhenium, iridium, germanium, or other such metals. They are supported on alumina and usually an acidic promoter such as chloride is added. The catalysts are called dual functional or bifunctional or multifunctional, because the metal, the support, and the promoter all play active roles in one or more of the reactions.

Since the first commercial catalytic reformer went on stream in 1940, hundreds of articles have been written describing new developments in reforming catalyst or technology. This may well be the first attempt to gather into a single publication the many varied aspects of catalytic reforming.

This book has two main purposes. First, to introduce catalytic reforming to those who have no practical knowledge of the process. These persons will be interested in how catalytic reforming fits into refinery processing, how it works, and its present state of development. The second purpose of the book is to give the reforming technologist a summary of the latest technology in operation, performance, and troubleshooting.

This is not a text on the theory of catalytic reforming. No one will be able to design and build a reformer from reading this book. The material presented centers on practical problems of operating commercial catalytic reformers. There is usually a diversity of answers to a reforming question. This book attempts to set out guidelines to point the way to a search for solutions. There are many references to NPRA Q&A transcripts; there are others not included. Anyone with a reforming problem should make use of the referenced sources.

Acknowledgments

This book could be written only because so many people helped me along the way. First, there was the suggestion by G.L. Farrar for the need of a book on catalytic reforming. When the project was begun, I received valuable assistance from Kathryne Pile, Brenda Bridges, and others at PennWell Publishing Co.

From my 42 years' refining experience at Phillips Petroleum Co., I owe a great deal to many people. Acknowledgment must be made to Gordon Allen, Dayton Lawson, and Paul Waddill for inspiration and guidance during those early years when catalytic reforming was just beginning to show its potential. More recently, George Constantikes, Ed Nakayama, Mack Potts, Larry Lew, Marvin Johnson, and Dean Montgomery were always willing to discuss reforming technology.

I cannot overlook the Phillips plant engineers who willingly answered my numerous requests for data from operating units. I am grateful to Dick Robinson, for wtihout his permission to peruse data on Phillips commercial reforming units the book would fall far short of its objective.

To Herb Bruch and Ray Whitson I express my thanks for the opportunity of serving those many years on the NPRA Q&A screening committee and especially to Herb for permission to use the information from NPRA publications. To all the people on the Q&A screening committees who shared with me their experiences in catalytic reforming, I am indeed grateful.

I am indebted to Frank Ciapetta, Robert Davidson, Vladimir Haensel, Alex Oblad, Marshall Sittig, and Wayne Warren for historical data from their 1950s articles.[1-5] I added information later than 1960. Any errors or omissions are unintentional.

[1] Ciapetta, Frank G. "Catalytic Reforming." *Petro/Chem Engineer* (May 1961): C–19.
[2] Davidson, Robert L. "Catalytic Reforming—From 1940 to Today." *Petroleum Processing* (August 1955): 1175.
[3] Haensel, Vladimir. "Platforming." *Petroleum Processing* (April 1950): 356.
[4] Oblad, Alex. "Catalyses—Its Status Today and Its Promise For Tomorrow." *Oil & Gas Journal* (21 March 1955): 182.
[5] Sittig, Marshall, and Wayne Warren. "How to Get Those Top Octanes." *Petroleum Refiner* (September 1955): 231.

Introduction

The first catalytic reforming unit went on stream November 1940 at the Pan American Refining Corp. (now American Oil Co.) refinery at Texas City, Texas.[1] Hydroforming, the name of the new technology, was a fixed-bed cyclic unit with two reactors on regeneration and two on process. The catalyst was about 9 wt % molybdenum oxide on activated alumina granules or pellets.

To regress a little, catalytic reforming was a logical outcome of catalytic developments in petroleum processing in the 1930s. Thermal cracking of naphtha, gas oil, and residuum dominated refinery processing in the early 1930s. Houdry catalytic cracking, a fixed-bed catalytic cracking process, conceived and developed by Eugene J. Houdry, went on stream in 1937 at the Sun Oil Co. refinery in Marcus Hook, Pennsylvania.[2]

Petroleum technologists no doubt saw the potential for the similar application of catalytic processing as a replacement for thermal cracking of naphtha. The incentive was higher-octane-number gasoline at increased yields. (At the time there was a push for higher-octane motor fuel.)

The article on Houdry catalytic cracking indicates Sun Oil was receptive to the Houdry process because Sun was one of the few companies not using tetraethyl lead (TEL) as octane-number booster, and they were concerned about meeting competition. At the same time, some companies foresaw that, in the event of war, there would be a far greater demand for high-octane aviation gasoline than thermal cracking of naphtha could produce.

---

[1] Davidson, op. cit.
[2] Oblad, Alex G. "The Contributions of Eugene J. Houdry to the Development of Catalytic Cracking." In *Heterogeneous Catalysis,* edited by B.H. Davis and W.P. Hettinger Jr., 61. Washington: American Chemical Society, Symposium Series 222, 1983.

xiv INTRODUCTION

Hydroforming was the only catalytic reforming process in use during World War II. Three units were built before the war, and five more were constructed during the war. Catalytic reforming made an important contribution to the war effort, producing toluene for trinitrotoluene (TNT, for explosives) and for aviation gasoline. Cycloversion, a process developed by Phillips Petroleum Co. about 1940, raised the octane number of straight-run naphtha by desulfurization over a bauxite catalyst. About 1943, the units were made cyclic, to combine desulfurization and reforming. Reforming was mild. Octane number gain was generally 7–12 research octane numbers (RON) clear and 12–20 RON + 3 ml TEL.

The demand for aviation gasoline drastically decreased after the war, and there was very little market for high-octane motor fuel. In the four-year period following the war, no new catalytic reformers were built. However, the automotive industry left no doubt that new postwar cars were designed for higher-octane gasoline. Catalytic reforming was the next step in modernizing refineries.

Platforming (a combination of platinum and reforming) was announced by Universal Oil Products Co. (now UOP Inc.) in March 1949. This process was the first commercial reforming technology to use a catalyst containing platinum. Platformers did not use a swing reactor like the cyclic reformers. The first units did not regenerate. The platinum catalyst operated about 9–18 months, then was removed and replaced with new catalyst. The usual reason for replacement was loss of reformate yield.

Platforming technology was first used commercially in October 1949 on a revamped thermal reformer at Old Dutch Refining Co.'s refinery in Muskegon, Michigan.[3,4] Three hours after start-up, the unit went down because of high reactor-shell temperature. The problem was soon solved, and the unit made about a nine-month run.

As it turned out, catalytic reforming was just getting a good start. Within five years of the announcement of the UOP process, nine new catalytic reforming processes were introduced to the petroleum industry. A chronological development of catalytic reforming is shown in Table I–1.

This book contrasts the design and operation of catalytic reformers of the 1950s with those of 1984. Catalytic reforming has come a long

---

[3]Haensel, Vladimir. "The Platforming Process—Personal Recollections." In *Heterogeneous Catalysis*, edited by B.H. Davis and W.P. Hettinger Jr., 141. Washington: American Chemical Society Symposium Series 222, 1983.
[4]Haensel, Vladimir. "Platforming." In *Petroleum Processing*, p. 356, April 1950.

INTRODUCTION xv

TABLE I-1
Catalytic reforming processes

Process	Developed by[a]	Date Process Announced	Date of First Commercial Use	Type of Process
Fixed-bed Hydroforming	Standard Oil Development Co., M. W. Kellogg Co., and Standard Oil Co. (Indiana)	1939	March 1940	Cyclic
Platforming	Universal Oil Products Co.	March 1949	October 1949	Semiregenerative
Catforming	Atlantic Refining Co.	February 1951	August 1952	Semiregenerative
Houdriforming	Houdry Process Corp.	May 1951	November 1953	Semiregenerative
Thermofor Catalytic Reforming	Socony-Vacuum Oil Co. Inc.	May 1951	March 1955	Moving-bed
Fluid Hydroforming	Standard Oil Development Co. and M. W. Kellogg Co.	May 1951	December 1952	Fluid Bed
Hyperforming	Union Oil Co. of California	February 1952	May 1955	Moving-bed
Orthoforming	M. W. Kellogg Co.	July 1953	April 1955	Fluid Bed
Ultraforming	Standard Oil Co. (Indiana)	November 1953	May 1954	Semiregenerative
Sovaforming	Socony-Vacuum Oil Co.	January 1954	November 1954	Semiregenerative
Powerforming	ESSO Research and Engineering Co.	March 1956	1956	Cyclic or Semiregenerative
Magnaforming	Engelhard and Atlantic Richfield Co.	1965	May 1967	Semiregenerative
IFP Reforming[b]	Institut Francais du Petrole	1960	1961	Semiregenerative
Rheniforming	Chevron Research Co.	1967	January 1970	Semiregenerative
Continuous Catalyst Regeneration Platforming	UOP Process Div. of UOP Inc.	1971	January 1971	Moving-bed

Notes: Phillips Petroleum Co.'s perco cycloversion process was on stream in 1940, raising octane number of straight-run naphtha primarily by desulfurization. By 1944 the process, using bauxite as catalyst, was expanded to a cyclic unit, raising octane number by combined desulfurization and reforming.
The above list does not include combination processes, that is, catalytic reforming plus aromatic extraction.
This tabulation does not include announcements of better catalysts. For example, the RD-150 catalyst of Baker and Co. with Sinclair Refining Co. (1954), or the new bimetallic catalysts announced from 1970 to 1980.

[a] Process developers are identified by company name at time of the process's announcement or its going into commercial service. Some companies have since changed names or merged with others. Companies offering processes for license in 1984 (Appendix V) are reviewed in chapter 7; their addresses are listed in Appendix VI.
[b] IFP continuously regenerative process was announced in 1968. The first unit was licensed in Italy in 1968.

way since 1940. Table I–2 illustrates the growth in installed U.S. reforming capacity from 1955 to 1984. The notable acceptance of bimetallic catalysts, since their appearance about 1970, is also indicated in Table I–2.

TABLE I–2
U.S. catalytic reforming capacity

January 1 of Year	Catalytic Reforming Capacity,[a] b/sd	Capacity Using Bimetallic Catalyst,[b] %
1955	569,330	—
1960	1,912,840	—
1965	2,063,570	0
1970	2,776,220	—
1972	3,169,130	30
1975	3,461,960	44
1977	3,670,210	60
1980	3,924,470	—
1982	3,978,180	80
1983	3,880,630	—
1984	3,862,800	—

[a]Cantrell, A. Annual refining issues of *Oil & Gas Journal (OGJ)*.
[b]Aalund, L.R. *OGJ* Annual Refining Issues, 1972–1984.

The pace for new catalytic reforming developments has not slackened; it seems to be intensifying. It appears inevitable that lead will be taken out of gasoline. To meet the resulting challenge for higher-octane-number reformate, many existing reformers will need better catalysts. Some older units will be replaced with new ones. It does, in fact, appear that refiners worldwide in the late 1980s and into the 1990s will be returning to what Sun Oil Co. attempted in the late 1930s when they began Houdry catalytic cracking, to make all gasoline unleaded.

Nomenclature and Units

NOMENCLATURE

ABD	average bulk density	MON0	motor octane number unleaded (clear)
API	American Petroleum Institute	MTBE	methyltertiarybutylether
°API gravity = $\dfrac{141.5}{\text{specific gravity}} - 131.5$		N + 2A	naphthene plus two times the aromatic content, vol %
		Ni-Mo	nickel-molybdenum
ASTM	American Society for Testing and Materials	NPRA	National Petroleum Refiners' Association
BTX	benzene, toluene, xylene		
CCR	continuous catalyst regeneration	PNA	paraffins, naphthenes, aromatics
CP	specific heat	PONA	paraffins, olefins, naphthenes, aromatics
Co-Mo	cobalt-molybdenum		
COP	catalyst-oriented packing	Q&A	question-and-answer session
Delta T	change in temperature		
EP	end point	$\dfrac{R + M}{2}$	(research octane number + motor octane number) ÷ 2
EPA	Environmental Protection Agency		
FCC	fluid catalytic cracking (also fluid catalytic cracker)	RON	research octane number
		RON0	research octane number unleaded (clear)
FEP	final end point	Rvp	Reid vapor pressure
HDN	hydrodenitrification	TEL	tetraethyl lead
HDS	hydrodesulfurization	TML	tetramethyl lead
H:HC	hydrogen-to-hydrocarbon ratio	TMP	trimethylpentane
		UOP Inc.	formerly Universal Oil Products Inc.
IBP	initial boiling point		
K	characterization factor	WABT	weighted average bed temperature
LHSV	liquid hourly space velocity		
		WAIT	weighted average inlet temperature
LPG	liquefied petroleum gas		
MCP	methylcyclopentane	WHSV	weight hourly space velocity
MON	motor octane number		

UNITS

Å	angstrom	MM	million, 10^6
bbl	barrel(s)	mol %	molar percentage
b/d	barrels per day	mol wt	molecular weight
b/sd	barrels per stream day	nl:l	normal liters:liter
BTU	British thermal units	ppb	parts per billion
cc	cubic centimeter(s)	ppm	parts per million
g	gram(s)	psi	pounds per square inch
gal	gallon(s)	psig	pounds per square inch gauge
hr	hour(s)	scf	standard cubic feet
kg	kilogram(s)	sec	second(s)
kw-hr	kilowatt-hour(s)	sq	square
l	liter(s)	ΔT	change in temperature
lb	pound(s)	vol %	volume percent
m	meter(s)	wt %	weight percent
ml	milliliter(s)		

To Leatrice, whose encouragement, assistance, and support made this book possible

CHAPTER 1

Catalytic Reforming in Refinery Processing

Catalytic reforming is a refining process that uses selected operating conditions and selected catalysts to convert naphthenes and paraffins to aromatics and isoparaffins. Hydrocarbon molecules are predominantly rearranged without altering the number of carbon atoms in the molecule, as described in chapter 2.

PRODUCTS FROM REFORMING

Only two reasons justify catalytic reforming in a petroleum refinery: to produce high-octane blending stock for motor fuel and to produce high-value aromatic hydrocarbons such as benzene, toluene, and xylenes (BTX).

Hydrogen is a valuable reforming product and, in most refineries, 80–1,500 standard cubic feet per barrel (scf/bbl) hydrogen yield is the source of part or all of the hydrogen used for hydrotreating, hydrocracking, or hydrorefining processes. However, catalytic reforming of naphtha cannot economically compete with steam-methane reforming and steam-naphtha reforming if hydrogen is to be the primary product. Hydrogen is more economically made by steam reforming than by cat reforming. Although the value of hydrogen yield may be included in the economics, no known catalytic reformer has been justified solely for hydrogen production. However, production of gasoline and aromatics makes cat reforming an economically viable process. The remainder of this book defines this process.

A reformer that is operated to make high-octane blending stock is a *motor fuel reformer*. When operated to make aromatics, it is a *BTX reformer*. The primary product from reforming is called *reformate*. Because reformer feedstocks contain paraffins, a BTX reformer yields a mixture of aromatics and paraffins; further processing such as extraction is necessary to recover aromatics of marketable purity. After removal of aromatics from the reformate, the remainder is called *raffinate*. In some refineries, a catalytic reformer may operate part time as a motor fuel unit and part time as a BTX unit. The position of the catalytic reforming unit (or cat reformer) in a typical fuels refinery processing scheme is shown in Fig. 1–1.

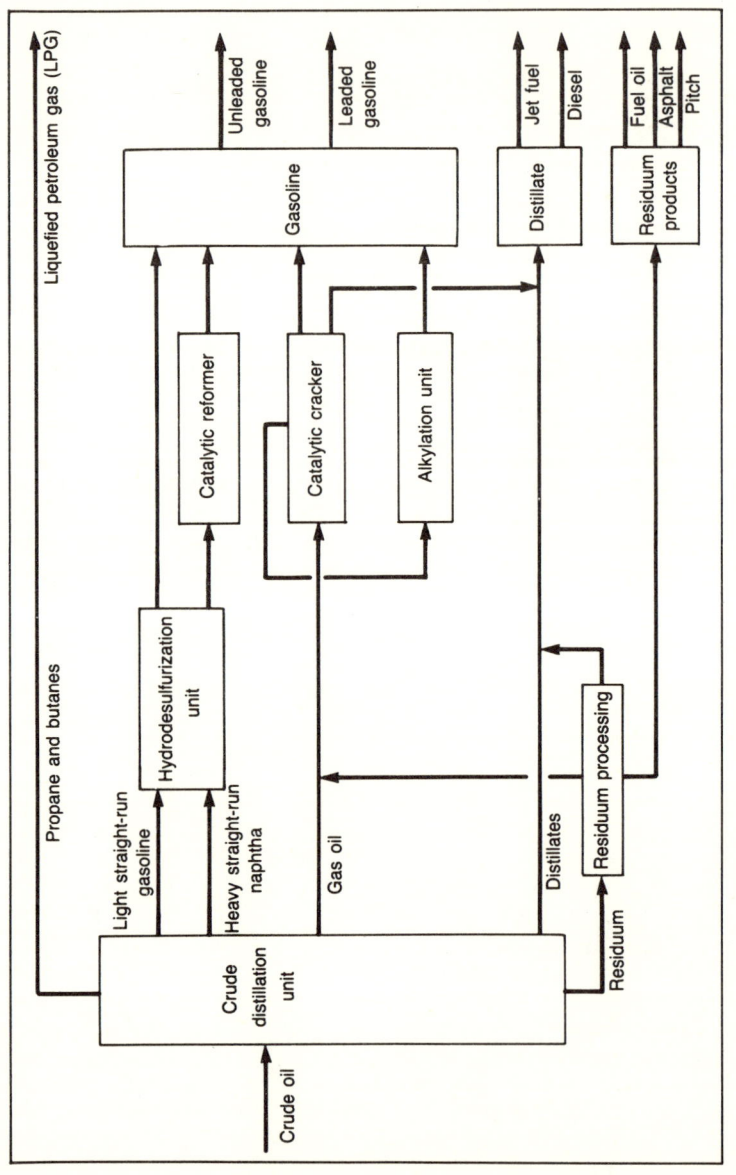

Fig. 1-1 Motor fuel refinery with catalytic reforming

FEEDSTOCKS FOR REFORMING

Because gasoline and distillates are usually the higher-priced refinery products, refinery processing units are designed to maximize conversion of crude oil to these products. Two of the lighter (lower boiling) streams from crude distillation are light straight-run (lt. st. run) naphtha and heavy straight-run (hvy. st. run) naphtha. Both these streams are in the gasoline boiling range (about 80–400°F). When the stream goes directly to motor fuel, it is called *gasoline* in refinery jargon. If the stream (that is, the feedstock) is charged to a process unit, it is generally called *naphtha*.

In Fig. 1–1, the light straight-run (80–180°F boiling range) fraction is hydrodesulfurized and blended directly to gasoline. The pentanes and hexanes in this boiling range are not converted to the desired aromatics in a cat reformer because they are not good cat reforming feedstock. The heavy straight-run naphtha is hydrodesulfurized and then charged to the cat reformer. The product from the cat reformer is blended with other refinery stocks to make finished gasoline.

Thus, the feedstock to the cat reformer from the crude unit is gasoline boiling range hydrocarbons, which were in the raw crude. These were separated from other crude hydrocarbons by fractionation. The gasoline (hvy. st. run naphtha), obtained directly from crude oil, is almost always of such low octane number that a refiner cannot use it in motor fuel. The other refinery blending stocks are usually not high enough in octane number to raise the motor fuel blends to the required octane number specification. So the purpose of the cat reformer is to raise the octane number of gasoline to a higher octane number.

As shown in chapter 5, if more high octane stocks are needed, catalytic reforming can raise the octane number of gasoline from catalytic crackers, hydrocrackers, cokers, and thermal crackers, as well as that of straight-run gasoline from the crude unit. Although feedstocks to a cat reformer are in the gasoline boiling range, they are usually called naphtha. For example, heavy straight-run gasoline from the crude unit is heavy straight-run naphtha; cat-cracked gasoline is cat-cracked naphtha; coker gasoline is coker naphtha.

In one respect cat reformers are unique in refinery processing: Gasoline is the feedstock as well as the desired product. Cat crackers, cokers, and thermal crackers all produce gasoline, but the charge stocks to them are higher-boiling hydrocarbons such as residuum and gas oil (a refined fraction of crude heavier than kerosene). Alkylation, polymerization, and isomerization units also produce gasoline, but they process lower-boiling-point hydrocarbons such as propylene, butenes, butanes, pentanes, and hexanes.

Cat reformers are valuable units because they can produce a wide range of octane number reformates, the product from the cat reformer,

4 CATALYTIC REFORMING

and very high octane number reformates. Reformates of 105 research octane number unleaded or higher (see next section) have been made on commercial units.* Catalytic reformers are used extensively to increase octane number of gasoline, one measure of gasoline quality. Therefore, a review of octane number relative to gasoline quality is in order.

OCTANE NUMBERS

To appreciate the significance of catalytic reforming in refinery operations, the use of *octane number* as a standard of gasoline quality must be understood. Octane rating has been used for years to measure the antiknock performance of gasoline. The higher the octane number, the less the tendency for a gasoline to produce a knocking sound in an automobile engine.

In 1923, a standard was established for measuring octane number of gasoline. The straight-chain paraffin n-heptane was assigned an octane number of zero, and a branched-chain paraffin, iso-octane (2,2,4-trimethylpentane), was assigned an octane number of 100. The octane number of a gasoline is determined by comparing its antiknock engine performance with various blends of n-heptane and iso-octane under specified laboratory conditions. Automotive engineers set compression ratios for particular engine designs. Engines with higher compression ratios require higher-octane-number fuel than those with lower ratios.

Two methods of determining motor fuel octane number are now in use:
1) the *research method*, ASTM† D–2699, a laboratory simulation of engine performance at low speed (reported as RON, research octane number)
2) the *motor method*, ASTMD–2700, a laboratory simulation of engine performance at high speed (reported as MON, motor octane number)

Road testing a number of different autos under varying conditions and gasolines has shown that the average of the RON and MON, (R + M)/2, gives an acceptable number for rating gasolines. This average is now a specification on gasoline and is the octane number displayed on the pumps at service stations.

Until 1974, almost all gasoline contained tetraethyl lead additive as an octane booster. With the advent of catalytic converters on most 1975

---

*Note: F–1 and RON clear both mean "research octane number unleaded." RON0 is a term used interchangeably with F–1 and RON clear, meaning RON with zero lead.
†ASTM stands for American Society for Testing and Materials

model U.S. cars in fall 1974, unleaded gasoline appeared, followed by lead phasedown in leaded gasolines. By 1984, 60–70% of the gasoline sold in the U.S. was unleaded. By 1990, an expected 80–90% of sales will be unleaded.[1,2]

A few examples of octane numbers of individual hydrocarbons and some selected refinery motor fuel blend stocks are shown in Table 1–1. Note that C_5^+ reformate (pentanes and heavier) from a reformer is the only gasoline stock that varies in octane number. Once the feedstocks are fixed, octane number of cat-cracked gasoline and alkylate is difficult to change by more than two to four units. Reformate octane number can be varied from 1 to 25 or more numbers. That flexibility is what makes the catalytic reformer so useful to a petroleum refiner.

Refinery Pool Octane Numbers

Catalytic reformers raise octane number, but the question each refiner must answer is "How much octane number do I need?" The higher the octane number, the lower the gasoline yield from reforming. Ulti-

TABLE 1–1
Octane numbers of selected hydrocarbons and refinery blend stocks[a]

	Research, ml TEL/gal[b]		Motor, ml TEL/gal		Octane Rating, (R + M)/2
	0.0	3.0	0.0	3.0	
n-Butane	94.0	104.0	89.0	104.7	91.5
i-Butane	102.0	118.0	97.0	—	99.5
n-Pentane	61.8	84.6	83.2	84.8	72.5
i-Pentane	93.0	104.9	89.7	107.3	91.4
n-Heptane	0.0	42.0	0.0	48.0	0.0
n-Octane	—	24.8	—	28.1	—
2,2,4-TMP[c]	100.0	115.5	100.0	115.5	100.0
Cyclohexane	84.0	96.6	77.6	87.4	80.8
Toluene	119.7	120.3	109.1	113.3	114.4
Lt. st. run	74.0	92.0	73.0	93.0	73.5
Cat cracked	90.0	97.0	80.0	85.0	85.0
Alkylate	93.0	104.0	92.0	106.0	91.0
C_5^+ reformate	90.0	98.0	81.0	89.0	85.5
C_5^+ reformate	95.0	101.0	85.0	93.0	90.0
C_5^+ reformate	100.0	104.5	90.0	94.0	95.0

[a] Octane reported for specific hydrocarbons may differ, depending on source. Octane number of refinery stocks will vary, depending on operation of each unit.
[b] To convert milliliters of tetraethyl lead per gallon (ml TEL/gal) to grams Pb/gal, multiply ml TEL/gal by 1.057.
[c] Trimethylpentane

6 CATALYTIC REFORMING

mately, the octane number needed depends on the proportion of the grades of gasoline to be sold.

Most service stations offer three gasoline choices: leaded regular (containing tetraethyl lead octane booster), unleaded regular, and unleaded premium. Leaded premium has nearly disappeared from the market because the low demand does not justify the cost of handling it. Leaded regular at the service station has an octane rating of 88–89 (R + M)/2 and contains up to 1.1 grams per gallon (g/gal) of lead. Unleaded regular is 87–88 (R + M)/2 octane rating. Unleaded premium is 91–92 (R + M)/2. A few unleaded premiums may be as high as 96–98 (R + M)/2. A refiner, by selectively blending gasoline stocks (for example, using stocks with high lead response in leaded regular), can minimize the octane number of reformate. However, such blending is complex and is usually done by computer. To illustrate catalytic reforming in refinery processing, we will use refinery total pool octane number. Total pool octane number is the volumetric average octane number—RON, MON, or (R + M)/2—of all refinery gasoline streams.

The projected gasoline grade distribution and refinery average clear (unleaded) pool octane ratings, (R + M)/2, through 1990 are shown in Figs. 1–2, 1–3, and 1–4. Using data from these figures, Table 1–2 shows

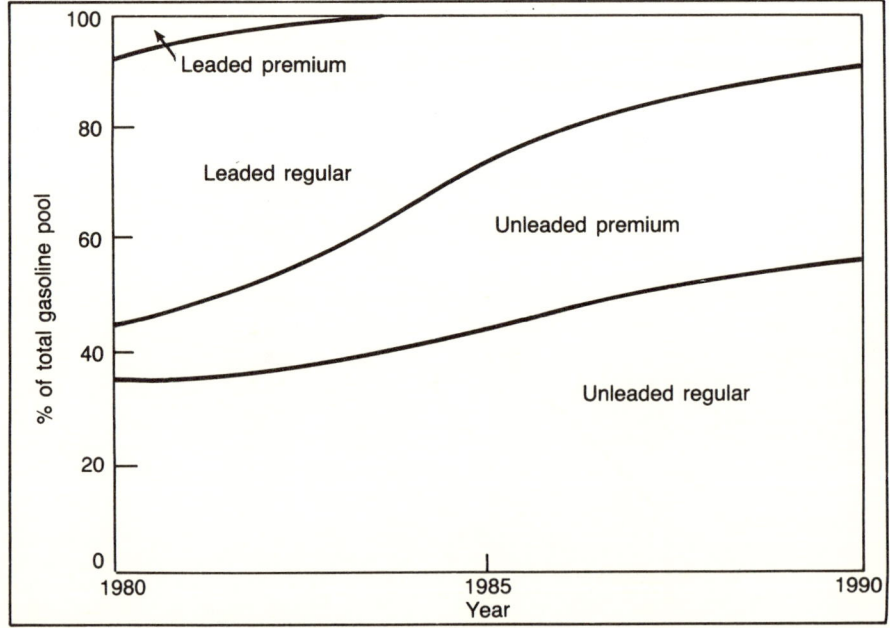

Fig. 1–2 Gasoline grade distribution *(after Lander, Hubbard, and Smith, ref. 2)*

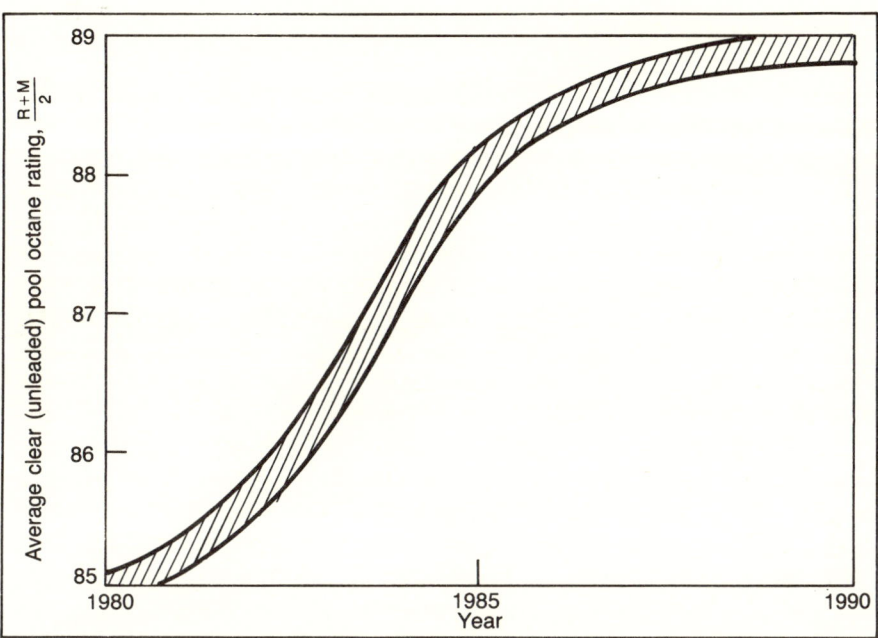

Fig. 1–3 Refinery average pool unleaded octane number demand *(after Lander, Hubbard, and Smith, ref. 2)*

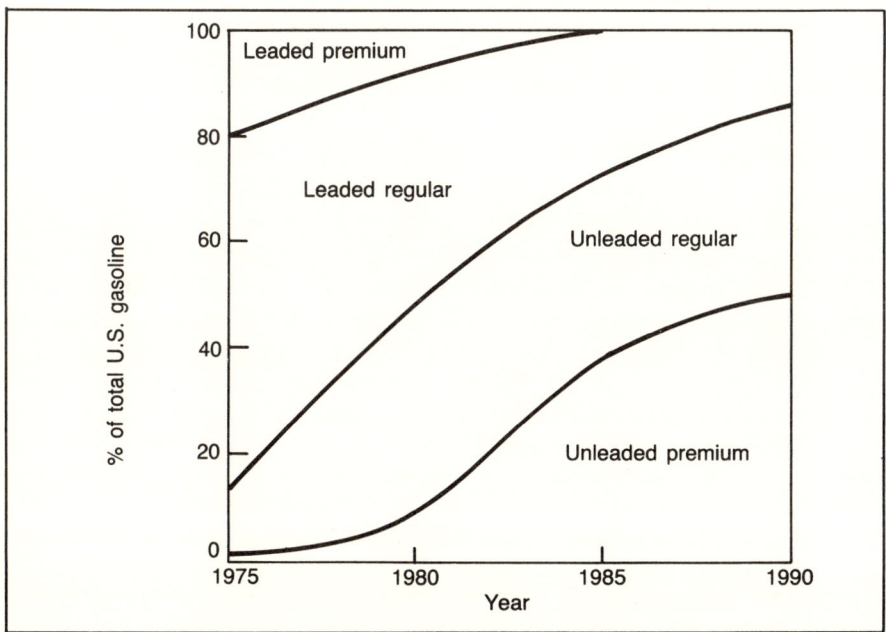

Fig. 1–4 How unleaded gasoline output will climb *(after Tippee, ref. 1)*

8 CATALYTIC REFORMING

TABLE 1–2
Refinery gasoline pool projected average unleaded octane number[a]

	Unleaded (R + M)/2	Gasoline Grade Distribution			
		1985		1990	
		Liquid Vol %[b]	Liquid Vol %[c]	Liquid Vol %[b]	Liquid Vol %[c]
Unleaded regular	87.5	36.0	42.0	35.0	55.0
Unleaded premium	91.5	36.0	33.0	50.0	37.0
Leaded regular	85.0	28.0	25.0	15.0	8.0
Average octane rating, (R + M)/2		88.2	88.2	89.1	88.8

[a]Does not reflect EPA's 1984 proposed lead reduction.
[b]23 June 1980 OGJ, p. 80.
[c]NPRA paper AM–83–50, Fig. 1.

that in 1985 the average refinery pool octane rating is slightly over 88. By 1990 this number will rise to about 89 because of increased unleaded gasoline production.

The Environmental Protection Agency (EPA) in 1984 announced plans to limit the lead content of leaded motor fuel to a maximum of 0.1 g/gal. If the EPA's proposals should become mandatory, the octane ratings for total refinery pool gasoline will increase 0.5–1.0 above those in Figs. 1–3 and 1–4.

Compare the pool octane number of the refinery depicted in Fig. 1–1 with and without reforming (Table 1–3) at 100,000 b/d crude charge. It is apparent that, without catalytic reforming, the octane number of the straight-run gasoline stocks is so low that the refinery gasoline pool octane rating of 74.2 is far below the 88 (R + M)/2 required for 1985 operation. It is also evident that reforming the heavy straight run to 100 RON clear (RON0) (95 octane rating) will raise the pool octane rating to 88, which will satisfy the 1985 requirement but not the 89 (R + M)/2 projected for 1990. In meeting octane specification, the refiner has taken a decrease of 6,600 b/d in gasoline production because 20% of the reformer charge was converted to hydrogen, methane, ethane, propane, butane, and coke in raising octane from 47 to 100 RON0.

These lower boiling products have some value to a refiner. Hydrogen is used for hydroprocessing, such as hydrodesulfurization, hydrodenitrification, demetallation, and aromatic saturation. Methane and ethane are used for fuel gas. Propane and butanes are sold for liquefied petro-

TABLE 1-3
Refinery gasoline pool without and with reforming

Blending Stock	Without Reforming				With Reforming			
	b/d	Vol %	RON0	MON0	b/d	Vol %	RON0	MON0
Lt. st. run	6,700	9.1	74	73	6,700	9.9	74	73
Hvy. st. run	22,000	29.7	47	46	0	—	—	—
Reformate	0	0.0	—	—	17,600	26.2	100	90
Alkylate	10,500	14.2	93	92	10,500	15.6	93	92
Cat cracked	28,000	37.8	90	80	28,000	41.5	90	80
n-Butane	6,800	9.2	94	89	4,600	6.8	94	89
Total gasoline pool	74,000	100.0	76.6	71.8	67,400	100.0	91.8	84.4
Reid vapor pressure[a] (Rvp), lb	10				10			
Average octane number, (R + M)/2	74.2				88.1			

[a]See chapter 3

leum gas (LPG) or are alkylated. Butanes can be used in gasoline to raise volatility.

By the year 1990 if the pool octane-number projections are correct, our refiner must do something to increase the refinery pool average clear octane number in the U.S. to about 89 (R + M)/2. Several possibilities may be considered:

1. Sell light straight-run gasoline
2. Reform light straight-run gasoline
3. Reform heavy straight-run gasoline to 100+ octane number
4. Isomerize the pentanes and hexanes in the light straight-run gasoline
5. Reform the cat-cracked gasoline
6. Produce methyltertiarybutylether (MTBE), a high-octane stock
7. Purchase high-octane stocks for blending

A discussion of all the above cannot be covered here; but of the options shown, only choices 1, 3, and 7 could be accomplished without a major capital investment. Reforming the heavy straight-run gasoline to an octane number higher than 100 RON0 might be possible, but reformate yield would drop rapidly as octane increased. Purchasing high-octane-number stocks puts the refiner in competition with others who want the same stocks and makes him dependent on outside sources.

Demand is increasing for light straight-run naphtha as feedstock to produce ethylene. The refiner might sell part, or all, of his light straight-run gasoline and realize a price of 80–90% that of gasoline. This case is shown in Table 1–4, where light straight-run gasoline is removed from the gasoline pool. The reformate is maintained at 100 RON0; the refin-

TABLE 1–4
Refinery gasoline pool without light straight run

Blending Stock	With Reforming			
	b/d	Vol %	RON0	MON0
Light straight run	0	—	—	—
Reformate	17,600	29.1	100	90
Alkylate	10,500	17.3	93	92
Cat cracked	28,000	46.3	90	80
n-Butane	4,400	7.3	94	89
Total gasoline pool	60,500	100.0	93.7	85.8
Rvp, lb	10			
Average octane rating, (R + M)/2	89.8			

ery average pool clear octane rating of 89.8 will meet the 1990 requirement. Little capital investment is required, and time will be gained for planning the next move if, in the future, a higher octane number is required.

Blending Octane Numbers

At this point in the discussion, note that individual stocks do not always blend to a final product that is proportionate to their octane numbers. That is, a 50:50 mixture of a 100-octane-number stock and an 80-octane-number stock generally will not produce a 90-octane-number blend. Every refiner knows this, and through tests has developed blending octane numbers that best fit his stocks. Thus the octane rating of a gasoline calculated from blending octane numbers will be very near the laboratory-determined octane rating. The following shows the calculated octane numbers and laboratory octane numbers from a reformate and alkylate blending study:

	Vol %	RON0	MON0	(R + M)/2
Reformate	50	93.0	82.6	87.8
Alkylate	50	90.5	89.5	90.0
Calculated blend	100	91.8	86.1	89.0
Actual blend	100	91.3	84.8	88.1

The actual blend was lower than the calculated average by almost one (R + M)/2. The octane numbers of reformate that would make the calculated blend equal to the actual blend can be derived by algebra. For example, the RON0 blending octane for reformate is computed as follows:

$$0.5x + 0.5(90.5) = 1.0(91.3)$$

$$x = \frac{91.3 - 0.5(90.5)}{0.5}$$

$$x = 92.1 \text{ RON0}$$

In this case, the octane numbers are 92.1 RON0 and 80.1 MON0, which are the blending octane numbers of the reformate. In the same way, blending octane numbers could be calculated for the alkylate. Blending octane number is customarily assigned to the stock that is the lowest percentage in the blend.

From the preceding example, we might conjecture that the difference between actual and calculated numbers could be attributed to the different composition of the two stocks. The reformate contains at least 40% aromatics, while the alkylate is mostly isoparaffins with no aromatics. But in this same study a 50:50 blend of alkylate and straight-run gasoline (neither of which contained aromatics) resulted in a blend of 0.5 higher octane rating than the calculated (R + M)/2. Further, a rather extensive study of blending octane numbers of aromatics was reported in a National Petroleum Refiners' Association (NPRA) paper at the 1980 annual meeting.[3] One conclusion was that the difference between blending values and octane number varies considerably among individual aromatics.

Because of the complexity of blending octane numbers, all blending in this book—unless otherwise noted—will be the calculated volumetric average.

CATALYTIC REFORMING FOR AROMATICS

Although catalytic reformers are most widely used for increasing octane number, many refiners find it economical to reform for certain aromatic hydrocarbons that can be sold as petrochemicals, usually at a much higher price than gasoline. These aromatics are benzene, toluene, and xylenes.

Cat reformers are particularly suited to yield these aromatics because refinery naphtha feedstocks contain a significant amount of naphthenes, which in the reformer are dehydrogenated to aromatics, and of paraffins, which are cyclized to aromatics (discussed in chapter 7). Operating conditions in reformer reactors are such that aromatics, once formed, are relatively stable and resist destruction or conversion to nonaromatics. A procedure for aromatics production follows.

The refinery process scheme of Fig. 1–1 must be changed to produce marketable aromatics. A new BTX catalytic reformer would be added. The installation of fractionation and extraction facilities would be required to separate and purify the benzene, toluene, and xylenes to specification-grade aromatics. The feedstock to the BTX reformer would likely be a straight-run naphtha. The refinery configuration would then be as shown in Fig. 1–5.

With this setup, the yield of aromatic hydrocarbons would be about 450 b/d benzene, 1,200 b/d toluene, and 800 b/d xylenes (C_8 aromatics). If these high-octane-number aromatics are taken out of the gasoline pool (Table 1–4), the average refinery pool octane number would drop from 89.8 to 88.9. The refiner could meet the 1985 gasoline pool octane number, but his pool octane number would be marginal by 1990.

SUMMARY

Catalytic reforming is a key process in a petroleum refinery. The catalytic reformer gives flexibility to meet gasoline octane number requirements. It can also make aromatics of high market value. Some refiners reform for octane number only; others produce aromatics, too.

CATALYTIC REFORMING IN REFINERY PROCESSING 13

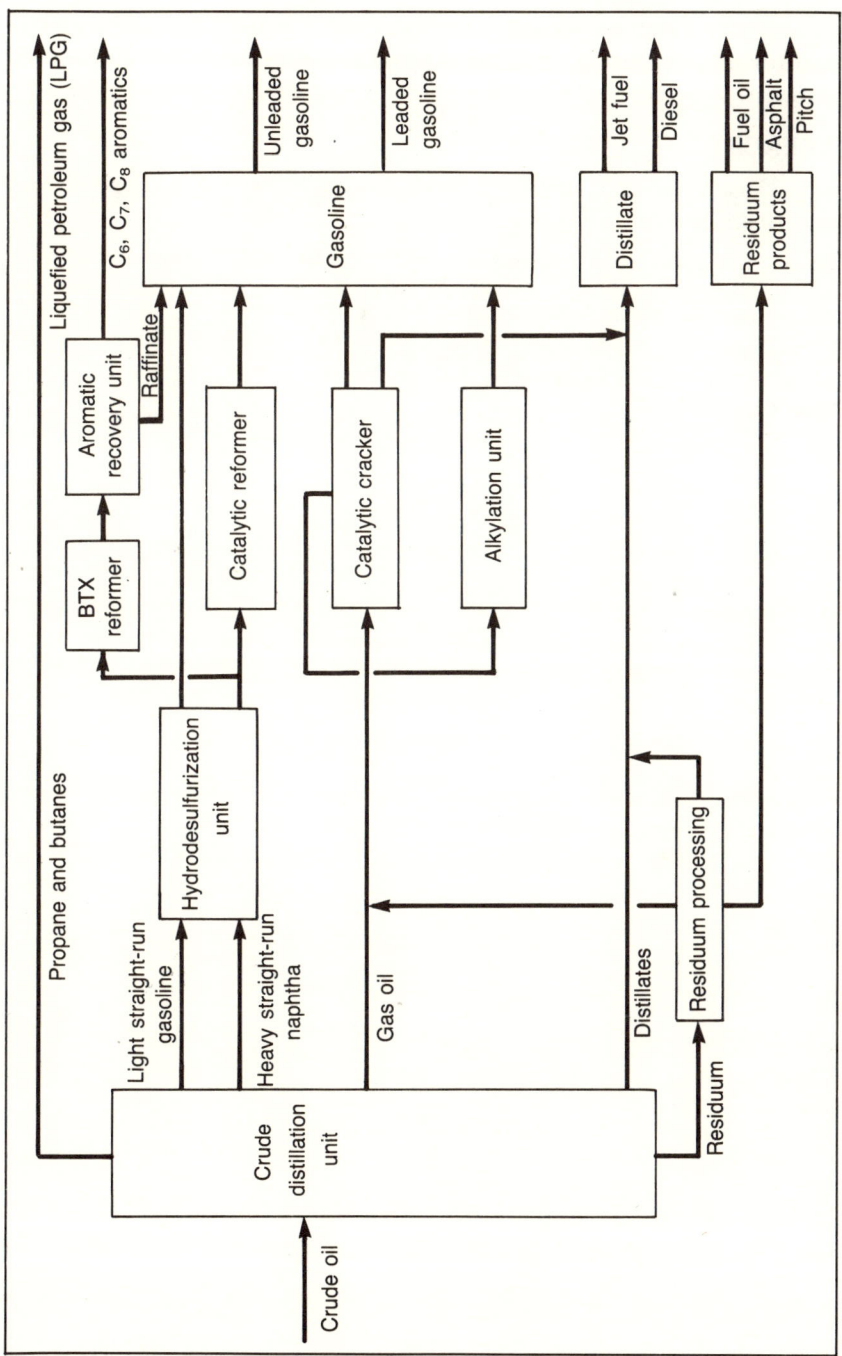

Fig. 1–5 Refinery with motor fuel and BTX catalytic reforming

Catalytic reforming additionally benefits the refiner by supplying hydrogen for hydrotreating, hydrocracking, and other hydrogen-consuming processes.

REFERENCES

1. Tippee, Bob. "U.S. Refiners Adjusting to Changing Requirements." *OGJ* 23 June 1980: 79–80.
2. Lander, E.P., J.N. Hubbard, and L.A. Smith. "The MTBE Solution: Octanes, Technology and Refinery Profitability." Paper AM–83–50 presented at the annual meeting of the NPRA, 1983.
3. Morris. W.E. "Octane Blending Effects of Aromatics." Paper AM–80–43 presented at the annual meeting of the NPRA, 1980 (contains data from API Project 45).

CHAPTER 2

How Catalytic Reformers Work— Chemical Reactions

As the name implies, reforming is carried out with the aid of a *catalyst*—the key to successful reforming. Fractionators, heaters, exchangers, and compressors are all parts of a reforming unit, but the reactors loaded with catalyst are the heart of the unit. Reforming catalysts are discussed in chapter 4. Let us first discuss the job the catalyst is expected to perform.

REFORMING REACTIONS

Until the advent of catalytic reforming, refining processes were nonselective (except a few like alkylation, polymerization, and isomerization)—that is, the feedstock was indiscriminately converted or cracked with little attempt to control specific chemical reactions. Catalytic reforming catalyst and operating conditions are designed to promote certain reactions and to inhibit others. Reactions most prevalent in catalytic reforming are dehydrogenation, isomerization, dehydrocyclization, and hydrocracking. The influences of catalyst composition, temperature, pressure, and space velocity on each reaction are of major importance (discussed in chapter 5).

Dehydrogenation

A principal reforming reaction is *dehydrogenation* because it produces aromatics that have high octane numbers or are valuable petrochemicals:

	Methylcyclohexane C_7H_{14}	Toluene C_7H_8	Hydrogen H_2
Mol	1.000	1.000	3.000
Mol wt	98.182	92.134	2.016

15

16 CATALYTIC REFORMING

	Methylcyclohexane C_7H_{14}	Toluene C_7H_8	Hydrogen H_2
lb	98.182	92.134	6.048
Wt %	100.000	93.840	6.160
lb/bbl	270.606	304.924	—
bbl	0.363	0.302	—
Vol %	100.000	83.200	—
RON0	73.8	119.7	—
RON + 3[a]	88.7	120.3	—
MON0	73.8	109.1	—
MON + 3	87.4	113.3	—
Scf/bbl	—	—	3959

[a] "+3" means "+3 ml of TEL/gal fluid."

Notice that conversion of a naphthene to an aromatic produces a 45.9 RON0 increase. Also note the decrease in volume and the increase in density, results of hydrogen loss. Thus, the maximum yield of toluene from methylcyclohexane is 83.2 vol %. This yield is seldom realized in commercial practice because theoretical equilibrium is not attained and because of ring opening reactions that convert the naphthene to a paraffin.

When a naphthene is converted to an aromatic, octane number is gained at a loss of volume. Some reforming reactions increase octane number with very little change in volume; an example is isomerization of paraffins.

Isomerization

Paraffins are isomerized at reforming conditions:

	C–C–C–C–C–C n-Hexane C_6H_{14}	⇌	C C | | C–C–C–C 2,2-Dimethylbutane (Neohexane) C_6H_{14}
Mol	1.000		1.000
Mol wt	86.172		86.172
lb	86.172		86.172
Wt %	100.000		100.000
lb/bbl	232.092		228.608
bbl	0.371		0.371
Vol %	100.000		100.020
RON0	24.8		91.8
RON + 3	65.3		106.0
MON0	26.0		93.4
MON + 3	65.2		113.1

Isomerization of a paraffin rearranges the molecule with essentially no change in volume but with a marked increase in octane number. In this example, the yield of neohexane was 100.02 vol % of the feed. The increase of 67 octane numbers would be a welcome change to any refiner needing higher-octane-number gasoline blend stocks. A refinery motor fuel pool containing just 2 vol % n-hexane would gain about 1.3 RON0 or 1.5 (R + M)/2 octane rating by isomerization.

Naphthene hydrocarbons may also isomerize in reforming reactors. When this happens, they almost immediately undergo dehydrogenation to an aromatic. A classic example of naphthene isomerization is conversion of methylcyclopentane (MCP) to benzene:

	MCP C_6H_{12}	Benzene C_6H_6	Hydrogen H_2
Mol	1.000	1.000	3.000
Mol wt	84.156	78.108	2.016
lb	84.156	78.108	6.048
Wt %	100.000	92.810	7.190
lb/bbl	263.508	309.343	—
bbl	0.319	0.253	—
Vol %	100.000	79.050	—
RON0	89.3	—	—
RON + 3	100.9	109.0	—
MON0	81.0	—	—
MON + 3	91.5	106.2	—
Scf/bbl	—	—	3,564

MCP is generally acknowledged to isomerize first to cyclohexane, which then dehydrogenates to benzene. Volume decreases, but octane number increases. Apparently this reaction involves ring opening, with a likelihood of forming paraffins as well as aromatics. On the other hand, paraffins can be cyclized to aromatics by what is known as dehydrocyclization.

Dehydrocyclization

Probably the most difficult reaction to promote in reforming is *dehydrocyclization*. It is a required reaction for production of a satisfactory yield of high-octane-number reformate. Most reforming feedstocks contain a substantial quantity of low-octane-number paraffins. The dehydrogenation and isomerization reactions of paraffins alone cannot produce enough aromatics and isoparaffins to yield high-octane-number reformate of 80 or higher RON0.

18 CATALYTIC REFORMING

A typical dehydrocyclization is:

C–C–C–C–C–C–C ⇌ Toluene + Hydrogen
n-Heptane   C₇H₈   H₂
C₇H₁₆

	n-Heptane C_7H_{16}	Toluene C_7H_8	Hydrogen H_2
Mol	1.000	1.000	4.000
Mol wt	100.198	92.134	2.016
lb	100.198	92.134	—
Wt %	100.000	91.950	—
lb/bbl	240.576	304.924	—
bbl	0.416	0.302	—
Vol %	100.000	72.560	—
RON0	0.0	119.7	—
RON + 3	41.9	120.3	—
MON0	0.0	109.1	—
MON + 3	48.1	113.3	—
Scf/bbl	—	—	3,644

As with naphthene isomerization, there are intermediate steps between the paraffin and the aromatic. The reaction is favored by high temperature and low pressure; it predominantly takes place in the last reactor in the train.

For cyclization, a paraffin with at least a six-carbon straight chain is needed. Consequently, n-pentane does not cyclize, nor does 2-methylpentane nor 2,2-dimethylpentane. However, n-hexane will cyclize to benzene. The longer straight-chain paraffins will dehydrocyclize more readily than the shorter chains.

Another reaction, looked upon with mixed emotion, that occurs in all commercial reformers is hydrocracking.

Hydrocracking

The breaking of the C–C bond in reforming operations is called *hydrocracking*. Hydrocracking can break a paraffin molecule into two molecules of lower molecular weight or can open the ring of a naphthene. Aromatics present in the feed or produced in the reactors do not normally undergo ring opening at reforming temperatures and pressures. Examples of hydrocracking reactions are:

C–C–C–C–C–C–C + H₂ → C–C–C + C–C–C–C
n-Heptane   Hydrogen   Propane   Butane
C₇H₁₆   H₂   C₃H₈   C₄H₁₀

	n-Heptane C_7H_{16}	Hydrogen H_2	Propane C_3H_8	Butane C_4H_{10}
Mol	1.000	1.000	1.000	1.000
Mol wt	100.198	2.016	44.094	58.120
lb	100.198	2.016	44.094	58.120
Wt %	100.000	—	44.010	58.000

lb/bbl	240.576	—	177.240	204.330
bbl	0.416	—	0.249	0.284
Vol %	100.000	—	59.740	68.280
RON0	0.0	—	—	—
RON + 3	41.9	—	—	—
MON0	0.0	—	—	—
MON + 3	48.1	—	—	—
Scf/bbl	—	911	—	—

The shearing of the C–C bond initially produces an olefin that is quickly hydrogenated. The C_5^+ reformate from a reformer contains only a trace of olefins, usually less than 0.5 vol %. These olefins are the result of hydrocracking close to the outlet of the last reactor where no time is left for hydrogenation. High temperature and high pressure accelerate hydrocracking. It usually takes place on the acidic sites of the catalyst but can also cause *pyrolysis* (chemical decomposition by heat) at temperatures above 1,100°F. Hydrocracking produces coke or coke precursors that cover active catalyst sites.

Hydrocracking is not all bad. For example, n-decane can split into n-heptane and propane. The n-heptane can then dehydrocyclize to toluene, a high-octane-number aromatic. Although a refiner usually likes to operate a reformer with a minimum amount of hydrocracking—to maximize C_5^+ output—a reformer can be operated at a higher than necessary octane to yield more propane and butanes. The price of LPG compared with the price of gasoline dictates this mode of operation.

Reaction Rates

All the above reactions occur in reforming, but all do not proceed at the same rate. The kinetics of reforming—important in designing and computer modeling reformers—is the subject of a number of publications (see references 1–10).

For the most part, an engineer assigned to a reforming unit will think more about *relative rate of reaction* of broad classes of hydrocarbons rather than the reaction rates of individual components. It is well known in reforming that naphthene dehydrogenation is a relatively fast reaction, is endothermic, and is favored by low hydrogen partial pressure.

The Atlantic Richfield NPRA paper AM–71–6 on reformer design[1] and the ESSO Research and Engineering paper on kinetic simulation modeling[2] both show that naphthene conversion in the first fraction (a few inches at the top) of the catalyst bed is fast compared to conversion of paraffins and aromatics. The appearance of substantial amounts of naphthenes, say, over 5 vol % in the reformate product, indicates poor catalyst activity or bypassing of the catalyst.

20　CATALYTIC REFORMING

A general characterization of relative reaction rates is:

Reaction	Relative Rate	Comments
Dehydrocyclization	1	Endothermic; promoted by low pressure and high temperature
Hydrocracking	4	Exothermic; promoted by high pressure and high temperature
Isomerization	12	Mildly exothermic; promoted by high temperature, only slightly affected by pressure
Naphthene dehydrogenation	100+	Endothermic; promoted by low pressure and high temperature

The above relative rates are broad generalizations. For example, the rate of dehydrocyclization of n-heptane is much faster than that of n-hexane. In general, the higher the number of carbon atoms per molecule, the more rapid the rate for the above reactions.

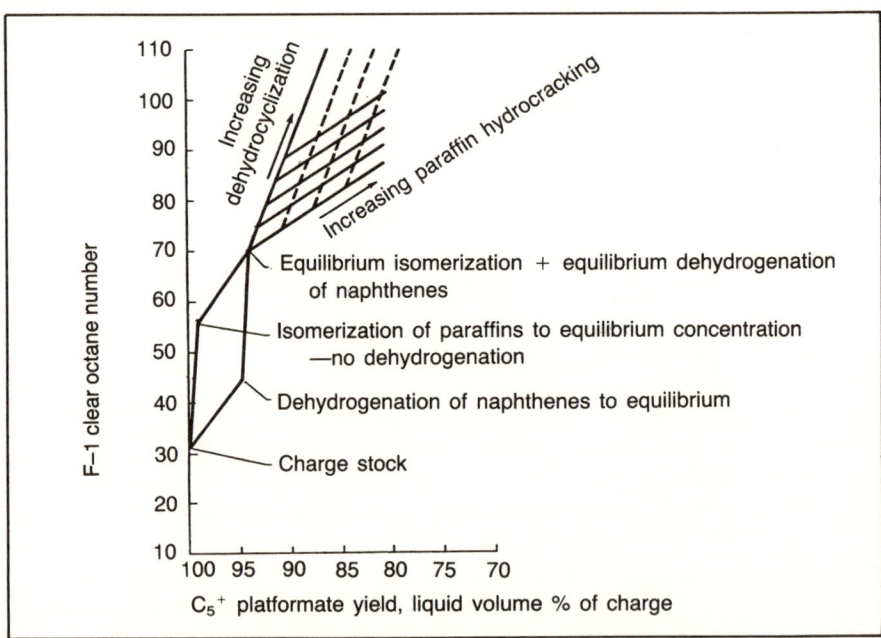

Fig. 2–1 Theoretical yield–octane study, Midcontinent-type stock *(courtesy UOP Process Div., after Haensel and Belden, ref. 8)*

HOW CATALYTIC REFORMERS WORK—CHEMICAL REACTIONS

A good illustration of the contribution of each of the reforming reactions to the octane number of reformate product is a theoretical yield–octane study by UOP (see Figs. 2–1 and 2–2).[8] The feedstock for Fig. 2–1 was assigned a composition of 49% paraffins, 44% naphthenes, and 7% aromatics to simulate a typical 200–400°F Midcontinent naphtha. The feedstock for Fig. 2–2 was assigned a higher percentage of paraffins and a smaller percentage of naphthenes to simulate a Mideast naphtha. These charts show the importance of dehydrocyclization and paraffin hydrocracking for raising octane number of reformer feedstock.

When dehydrogenation of naphthenes and isomerization of paraffins have reached equilibrium, the octane number of the reformate is only 60–70 RON clear. This number is far below the 90–100 required for unleaded gasoline. Paraffin hydrocracking increases octane number, but at the penalty of substantial yield loss. Dehydrocyclization produces the higher-octane-number reformate and better yield.

Advances in catalytic reforming technology have emphasized unit design and catalyst developments that promote the dehydrocyclization reaction.

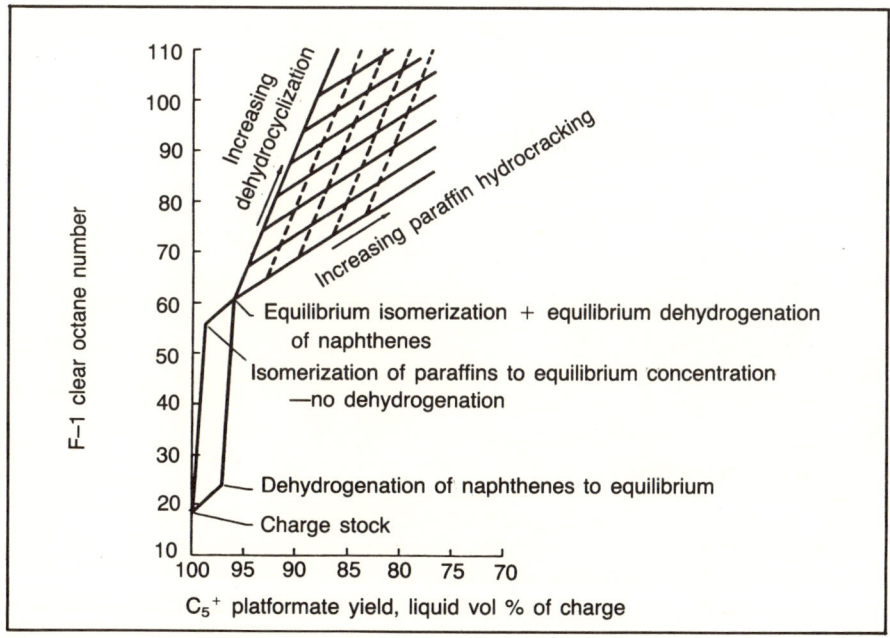

Fig. 2–2 Theoretical yield–octane study, Mideast-type stock *(courtesy UOP Process Div., after Haensel and Belden, ref. 8)*

Demethylation

One more reaction that can occur in reformer reactors, sometimes with disastrous results, should be mentioned. At very severe operating conditions of high pressure, high temperature, superactive catalyst, and low space velocities, *demethylation* occurs. One example of demethylation is:

C–C–C–C–C–C–C n-Heptane C_7H_{16}	+	$6H_2$ Hydrogen H_2	→	7C Methane CH_4
Mol 1		6		7

This reaction is highly exothermic, so it releases a great amount of heat.

The heat release for the reaction is calculated to be 154,000 BTU/lb mol of n-heptane converted. If it is not rapidly removed, the heat further promotes the reaction, ultimately melting the catalyst and sometimes the reactor steel wall.

In early reforming units demethylation seemed more likely to occur during start-up with fresh catalyst. The newer-generation catalysts, presulfiding the catalyst, and strict attention to start-up procedure have essentially eliminated demethylation during start-up. The problem is more likely to occur when reactors are operated over lengthy periods between internal inspections.

REACTIONS IN COMMERCIAL UNITS

If reformer feedstocks were limited to a few paraffins, naphthenes, and aromatics with well-known physical and chemical properties, estimating yield and octane numbers would not be difficult. In actual practice, reformer feedstocks may contain well over one hundred identifiable components, plus some unknowns and trace contaminants.

Tables 2–1 and 2–2 illustrate a chromatographic analysis—obtained during a special test—of a feed to a motor fuel reformer.

TABLE 2–1
Reformer charge analysis

	Wt %
n-Butane	0.01
Isopentane	0.07
n-Pentane	0.10
3M pentane	0.29
NC_6	0.26
2,2/2,4-DM pentane	0.59
2,2,3-TM butane	0.04

HOW CATALYTIC REFORMERS WORK—CHEMICAL REACTIONS 23

3,3-DM pentane	0.46
2M hexane	0.60
2,3-DM pentane	0.31
3M hexane	1.02
Cyclohexane	0.80
1,1-DMC pentane	0.13
3-Ethylpentane	0.10
cis 1,3-DMC pentane	0.45
trans 1,3-DMC pentane	0.44
trans 1,2-DMC pentane	0.51
n-Heptane	3.90
Benzene	0.27
2,5/2,4-DM hexane	0.84
cis 1,2-DMC pentane	0.36
Methylcyclohexane	10.04
1,1,3-TMC pentane	0.10
Ethylcyclopentane	0.80
3,3-DM hexane	0.57
2,3-DM hexane	0.99
3,4-DM hexane	0.44
2M heptane	3.58
4M heptane	0.89
3M heptane	2.95
1,C-2,T-4 TMC pentane	0.12
1,1-DMC hexane	3.52
C-1,3 DMC hexane	0.82
T-1,4 DMC hexane	0.36
n-Octane	8.43
1-Me-2ethyl CyC_5	0.04
Isopropyl CyC_5	1.32
1M,2ethyl CyC_5	0.10
Toluene	4.47
T-1,3 DMC hexane	0.32
C-1,2 DMC hexane	1.39
1,C-3,T-5 TMC hexane	0.60
Ethycyclohexane	2.35
1,1,3-TM cyclohexane	2.60
1,2,3-TM cyclohexane	0.13
2,2-DM heptane	0.39
2,3-DM heptane	0.60
2M octane	0.93
4M octane	1.78
3M octane	1.62
1,1-Diethyl CyC_5	0.24
1,3,5-TM CyC_6	0.18
2,3,5-TM hexane	0.86
2,3,4-TM hexane	0.74
2,2,5-TM hexane	0.40
n-Nonane	5.68
Ethylbenzene	4.38
M-/P-xylene	3.28
O-xylene	1.88
Isopropyl CyC_6	0.43
n-Propyl CyC_6	0.94
Tert butyl CyC_6	0.40
Cumene	1.08

TABLE 2–1
Reformer charge analysis—cont'd

	Wt %
C_9 naphthenes	0.05
C_{10} paraffins	2.69
n-Decane	0.53
p-Cymeme	0.49
C_{10} naphthenes	0.64
Isopropylbenzene	0.74
n-Propylbenzene	0.53
NC_{11}	1.70
C_{12} paraffins	1.53
1,3,5-TM benzene	0.67
1,2,4-TM benzene	0.94
1,2,3-TM benzene	0.22
M-/P-ethyltoluene	3.13
C_{12} naphthenes	0.44
C_{10} aromatics	1.41
	100.00
Paraffins	45.89
Naphthenes	30.62
Aromatics	23.49
	100.00

TABLE 2–2
Feedstock characteristics

Gravity	52.8°API
ASTM Distillation	
Initial boiling point (IBP)	219°F
10 vol %	245°F
30 vol %	259°F
50 vol %	277°F
70 vol %	299°F
90 vol %	338°F
End point (EP)	396°F

FEEDSTOCK CHARACTERIZATION

Most refiners are not equipped to run such a detailed analysis. Even if they were, day-to-day reforming operation would not justify such analysis.

The method used most by refiners to estimate catalyst performance (see Fig. 2–3) is the naphthenes-plus-aromatics in the feed, expressed as liquid vol % = (N + A) or liquid vol % = (N + 2A).

Common practice for licensors of catalytic reforming processes is to furnish correlations for estimating yields and octane numbers based on naphthene and aromatic content of the feed. Also, charts to estimate

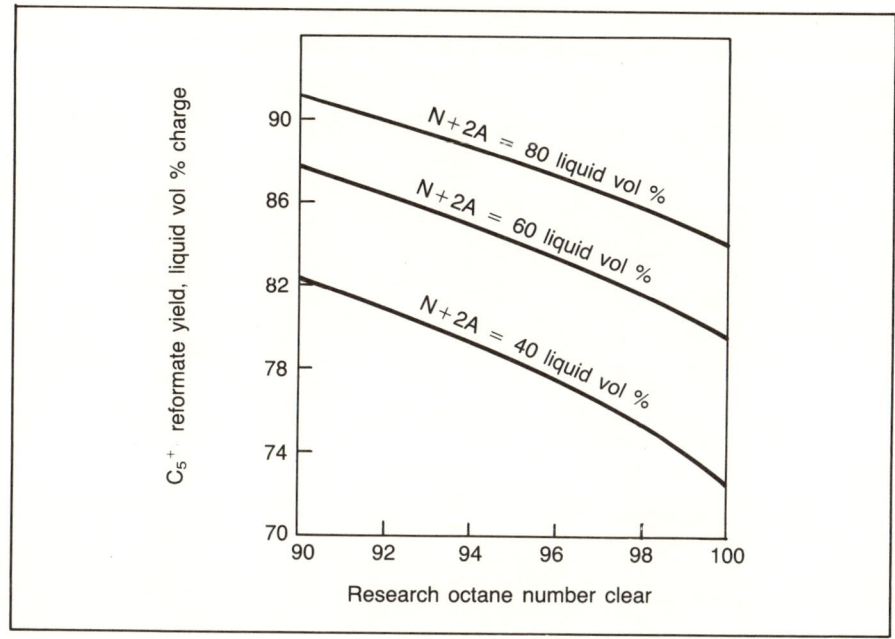

Fig. 2–3 Effect of charge N + 2A on yield *(courtesy UOP Process Div., after D'Auria, Tieman, and Antos, ref. 9)*

weighted average (reactor) inlet temperature *(WAIT)* or the weighted average bed temperature *(WABT)* and the liquid hourly space velocity *(LHSV)* or weight hourly space velocity *(WHSV)* required are supplied by licensors.

A typical motor fuel reformer feed and reformate analysis is:

	Feed	C_5^+ Reformate
Paraffins, vol %	57.6	53.2
Naphthenes, vol %	26.7	2.3
Aromatics, vol %	15.7	44.5
	100.0	100.0
N + 2A	58.1	—
RON0	—	94.3
RON0 + 3	—	100.6

The octane number is not usually run on the feedstock. In this case the feedstock is a Midcontinent naphtha, and the RON0 should be in the range of 40–70.

Since BTX reformers (operated specifically to produce certain aromatics) operate with lower end point feedstock, the feedstock analysis usually identifies more components than for a motor fuel reformer. Tables 2–3 and 2–4 record feedstock from one BTX reformer.

TABLE 2–3
Sample ASTM distillation (178 IBP to 327 FEP)

Compound	Vol %
Isopentane	0.12
n-Pentane	0.35
C_6 paraffins	14.60
Methylcyclopentane	3.14
Cyclohexane	3.73
Benzene	2.31
C_7 paraffins	11.38
C_7 naphthenes	13.00
C_7 aromatics	7.03
C_8 paraffins	13.10
C_8 naphthenes	9.20
C_8 aromatics	9.13
C_9 paraffins	7.12
C_9 naphthenes	4.89
C_9 aromatics	0.90
	100.00
Paraffins	46.69
Naphthenes	33.95
Aromatics	19.36
	100.00

TABLE 2–4
Yields of aromatics, based on reactor charge

Compound	Vol %
Benzene	6.8
C_7 aromatics	16.2
C_8 aromatics	19.6
C_9^+ aromatics	6.3

The C_5^+ reformate yield was 84.1 vol % of reactor charge and 94.0 RON0. The yield of aromatics can be varied, depending on reactor pressure and operating severity (RON0), as discussed in chapter 6.

Reforming reactions are well recognized in the industry, but this chapter shows that the extent to which they are monitored depends on the needs of individual plants.

SUMMARY

Dehydrogenation of naphthenes to aromatics, isomerization of n-paraffins to isoparaffins, and dehydrocyclization of paraffins to aromatics are the principal reactions of catalytic reforming. These reactions convert low-octane-number naphtha to high-octane-number gasoline stocks essential for production of unleaded gasoline. These same reactions convert selected naphthas charged to BTX reformers to high yields of aromatics such as benzene, toluene, orthoxylene, metaxylene, and paraxylene.

In commercial operations, reformer feedstocks contain many components. So it is not practical to use individual hydrocarbons to evaluate reformer performance. Refiners use yield–octane correlations based on the naphthene and aromatic content of the feedstock to evaluate motor fuel reformer performance. BTX reformer performance evaluation requires a more detailed breakdown of feedstock components. For example, naphthenes are classified by carbon number, i.e., C_7 naphthenes, C_8 naphthenes, and C_9 naphthenes. Paraffins and aromatics are classified in the same manner. The way a reformer feedstock is characterized is the choice of each refiner.

REFERENCES

1. Decker, W.H., W.E. Haynes, and M.H. Dalson. "Optimizing Reformer Design for High Severity Operation." Paper AM–71–6 presented at the annual meeting of the NPRA, 1971.
2. Kmak, W.S. and A.N. Stuckey Jr. "Powerforming Process Studies with a Kinetic Simulation Model." Paper 56a presented at the national meeting of American Institute of Chemical Engineers (AIChE), 1973.
3. Gates, Bruce C., James R. Katzer, and G.C.A. Schuit. *Chemistry of Catalytic Processes*. New York: McGraw-Hill, 1979.
4. Ciapetta, F.G., and D.N. Wallace. "Catalytic Naphtha Reforming." *Cat. Rev.* 5 (I), (1971): 67–158.
5. Krane, H.G., A.B. Groh, B.L. Schulman, and J.N. Sinfelt. "Reactions in Catalytic Reforming of Naphthas." Paper 4, Section III, presented at the Fifth World Petroleum Congress, 1959.
6. Pollitzer, E.L., J.C. Hayes, and V. Haensel. "The Chemistry of Aromatics Production Via Catalytic Reforming." Paper presented at the meeting of the American Chemical Society, New York City, September 1969.
7. Kugelman, Alan M. "What Affects Cat Reformer Yield." *Hydrocarbon Processing* January 1976:95.
8. Haensel, Vladimir, and D.H. Belden. "Platforming—a Progress Report." Paper presented at the 21st midyear meeting of the American Petroleum Institute's (API) Division of Refining, May 1956.
9. D'Auria, James, William C. Tieman, and George J. Antos. "Recent Platforming Catalyst Developments." Paper AM–80–49 presented at the annual meeting of the NPRA, 1980.
10. Jenkins, J.H., and T.W. Stephens. "Kinetics of Cat Reforming." *Hydrocarbon Processing* November 1980:163.

CHAPTER 3

How Catalytic Reformers Work— Process Design

The process design of a catalytic reforming unit follows the basic configuration of most refinery catalytic process units. The generalized flow scheme is:

Feed Preparation → Temp. Control → Reactor(s) → Product Recovery

Each component of the flow scheme is engineered to the needs of a particular process. For example, the reactor of a fluid cat cracker (FCC) differs in size, shape, and metallurgy from that of a catalytic reformer, but each is designed so desired reactions can take place.

The reaction section of an FCC unit rarely has more than one reactor. A catalytic reformer may have from three to five reactors in its reaction section. Variations within a unit are necessary, but the basic flow scheme is still recognizable.

Fig. 3–1 shows the basic elements of a catalytic reformer for a step-by-step description of the process. More details and variations in design are discussed in chapters 5 and 6.

FEED PREPARATION

The charge, or feedstock, to the cat reformer is a naphtha that has been processed in a feed-preparation unit to remove contaminants such as sulfur, nitrogen, arsenic, and lead. These contaminants are either temporary or permanent poisons to the catalyst and must be eliminated for satisfactory catalyst performance.

The feed-preparation unit itself is a catalytic unit using a cobalt-molybdenum (Co-Mo) or a nickel-molybdenum (Ni-Mo) catalyst. Hydrogen is included with the feed to the feed-preparation units. These feed-preparation units are called *hydrodesulfurization* (HDS) *units*, or *hydrotreaters*.

HOW CATALYTIC REFORMERS WORK—PROCESS DESIGN

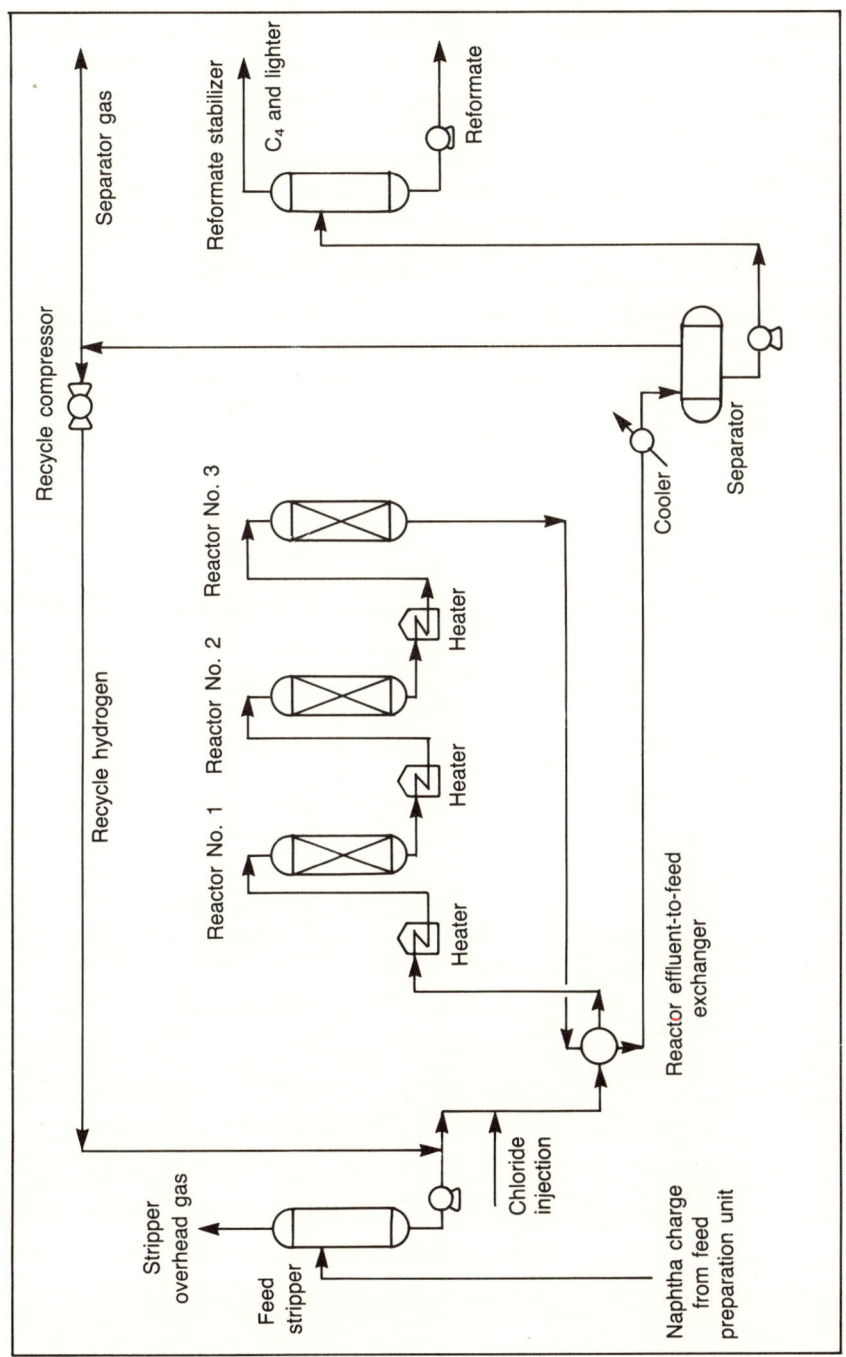

Fig. 3–1 Process flow scheme of a catalytic reformer

30   CATALYTIC REFORMING

After passing over the Co-Mo or Ni-Mo catalyst, the naphtha goes to a stripper or fractionator that removes the hydrogen sulfide, ammonia, water, and light hydrocarbons formed in the hydrotreater reactor. The feed stripper (Fig. 3–1) could be either on the tail end of the hydrotreater or on the front end of the reformer.

In the flow scheme (Fig. 3–1), the reformer naphtha charge is fractionated to the correct end point, probably at the crude unit. Only the initial boiling point is adjusted in the feed stripper.

A number of early reformers were designed with a centerwell in the prefractionator so both initial boiling point and end boiling point were made to specification in one vessel. Fig. 3–2 (trays not shown) illustrates a centerwell prefractionator and a fractionator with the reformer reactor feed made off the bottom.

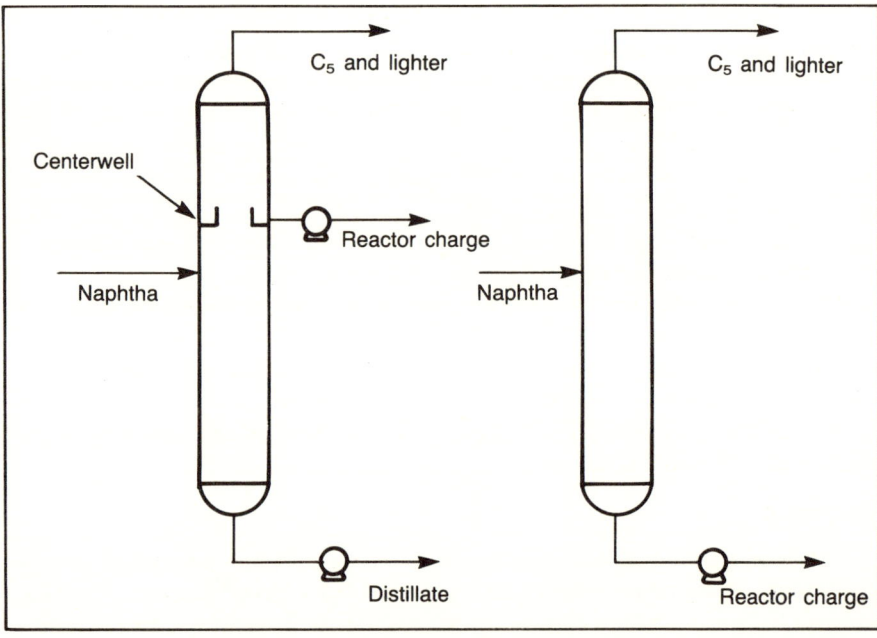

Fig. 3–2   Reformer feed prefractionators

The reactor feed is by far the greatest volume yield off the prefractionator. With a centerwell, almost all the naphtha must be vaporized. This requires much more energy than the scheme of yielding, or making, reactor charge off the bottom of the prefractionator.

On the other hand, some refiners like to include a light distillate, such as kerosene or jet fuel, in the feed to the hydrotreater for desulfurization, along with the reformer feed. In that case the centerwell design provides a means of separating the distillate from the reformer reactor charge.

The reformer reactor charge, pretreated and fractionated to proper initial and end boiling points, is combined with a gas stream containing 60–90 mol % hydrogen. Also, a chloride chemical representing 0.5–5.0 weight parts per million (ppm wt) of the reactor charge is injected. In reforming nomenclature, reactor charge means the naphtha feed. Total reactor charge is the naphtha feed plus the hydrogen-bearing recycle gas stream.

TEMPERATURE CONTROL

Now the reactor charge is ready for reforming, except that it must be raised to the proper temperature for the reforming reactions to occur when the charge contacts the catalyst. As shown in Fig. 3–1, total reactor charge is heated, at first by exchange with effluent from the last reactor, and is finally brought up to the No. 1 reactor inlet temperature in the charge heater. (Effluent is total vapor flowing out of the last reactor.)

The reactor effluent-to-feed exchanger is one key to energy conservation in a catalytic reformer. The reactor effluent, which may run at temperatures as high as 950–1,000°F, must be cooled to 90–125°F for flash separation—separation of vapor and liquid without fractionation—of hydrogen from the reformate.

The reactor effluent-to-feed exchanger is sometimes called the *combined feed exchanger*. This exchanger cools reactor effluent and, at the same time, adds heat to the incoming total reactor charge. In reformers built in the 1950s and 1960s, this heat exchange consists of several banks of heat exchangers, arranged for parallel flow. Newer units use one large single-pass vertical exchanger.

REACTORS

Catalytic reforming is a vapor-phase process. After passing through the reactor effluent-to-feed exchanger and the charge heater, the total reactor charge is 100% vapor, is up to reaction temperature, and is ready to contact the reforming catalyst.

The flow scheme (Fig. 3–1) shows three reactors. Reformers have been built with four or five reactors. Although the Fig. 3–1 flow sheet shows all reactors the same size, reactors are different sizes in many units, with the smallest reactor in the No. 1 position and the largest as the last reactor.

Some units originally built with three reactors have been revamped by adding one or more reactors. The additional reactors increase catalyst volume, which permits increased time between catalyst regenerations, increased charge rate, or higher-octane-number reformate. Engineers getting their first introductions to catalytic reforming frequently ask, "Why so many reactors? Why not use one large reactor instead of three or four small ones?" The answer lies in the nature of the reforming reactions (discussed in chapter 2).

The major reactions, such as dehydrogenation, are endothermic and very fast. The first contact with catalyst causes a rapid decrease in temperature. Very quickly the temperature in the reactor drops so low that the reaction rate becomes too slow for commercial operation. It is not unusual in the No. 1 reactor to have a temperature decrease between inlet and outlet of 180–280°F. For this reason, catalytic reformers are designed with multiple reactors and with heaters between reactors to maintain reaction temperatures at operable levels.

As total reactor charge passes through the reactor train (three or more reactors in series flow), the reactions become less and less endothermic and temperature differential across the reactors decreases. In fact, the last reactor may show zero endotherm or, in some cases, a small temperature increase.

The flow scheme of Fig. 3–1 shows a separate heater for each reactor. A number of units have heaters arranged this way, but there are also units with two or more heating coils in one large heater. The coils are separated within the heater by a bridgewall arrangement. A bridgewall is firebrick between heating coils.

PRODUCT RECOVERY

The effluent from the last reactor, at temperatures from 950–1,000°F, is cooled by heat exchange with the reactor charge, then further cooled by air and water exchangers to 100–125°F.

The stream then enters the separator vessel (called the *product separator*), where flash separation of hydrogen and some of the light hydrocarbons—primarily methane and ethane—takes place. The flashed vapor, containing 60–90 mol % hydrogen, passes to a compressor and then circulates to join the naphtha charge from the feed stripper. The hydrogen circulating to the naphtha charge is called *hydrogen recycle*. Excess hydrogen from the separator is yielded to fuel gas or to hydrogen-consuming units in the refinery, for example, hydrotreaters.

The separator liquid, comprised mostly of the desired reformate product but also containing hydrogen, methane, ethane, propane, and butanes, is pumped to the reformate stabilizer or fractionator. Reformate off the bottom of the stabilizer is sent to storage for gasoline blending or, in the case of a BTX reformer, to an aromatic-extraction unit. Some

HOW CATALYTIC REFORMERS WORK—PROCESS DESIGN 33

refiners operate the stabilizer to leave sufficient butane in the reformate to meet Reid vapor pressure (Rvp) specifications when making gasoline blends.

REFORMER CLASSIFICATIONS

The flow diagram of Fig. 3–1 is purposely simplified. This reformer can be classed as one of four types of reformers generally used in the petroleum industry. The four reformer classifications (discussed in chapter 7) are as follows.

1. *Semiregenerative.* This means the reformer processes feedstock for a time and then shuts down for regeneration and rejuvenation of the catalyst. The time between regenerations is called a *cycle* and is expressed in months. (A cycle is sometimes reported as barrels of feed per pound of catalyst.) Cycles vary from three months to three years. A cycle may be terminated for a number of reasons, but the most common is poor catalyst performance. This shows up as a loss of reformate yield or as too-high reactor-inlet temperatures approaching the maximum allowable for reactor metallurgical construction.

    The regeneration procedure, which includes rejuvenation, restores the catalyst to fresh-catalyst condition. The reformer, as a semiregenerative unit, will have a few connections not shown in Fig. 3–1 for regeneration purposes. These connections allow addition of air and nitrogen or oxygen and nitrogen, injection of hydrogen chloride, instrumentation, and neutralizing solution to be circulated (see chapter 5). The catalyst in a semiregenerative unit is expected to retain its usefulness over multiple regenerations and to have an ultimate life of 7–10 years or longer.

2. *Nonregenerative.* Some reformers do not regenerate catalyst because the cycle is long enough for the refiner to justify replacing—instead of regenerating—the catalyst. The reformer of Fig. 3–1 can easily operate in this manner.

3. *Cyclic.* The unique feature of a cyclic reformer is that it has a special valving-and-manifold system so any reactor can be isolated and regenerate catalyst while the other reactors are reforming. In contrast to a semiregenerative reformer, which requires a shutdown of the entire unit to regenerate catalyst, a cyclic unit can continue operation over long periods between shutdowns. Note, however, that a cyclic unit can operate as a semiregenerative reformer.

    Since the usual reason for regeneration is to burn off carbon that has fouled the catalyst, a cyclic unit can reform at lower pressure and higher severity than semiregenerative units, even though rapid coking results. The cycles (switching reactors in a cyclic unit) vary from

34 CATALYTIC REFORMING

a few hours to weeks or months. Although the reactors can be operated in any sequence, the last reactor is usually the one taken off-line for regeneration.

4. *Moving-bed or continuous catalyst regeneration.* These units, as the names imply, permit the catalyst to be moved continuously through the reactors, to be withdrawn from the last reactor, to be regenerated in a regeneration section, and to be returned to the first reactor as fresh catalyst. Compared to a fluid cat-cracking unit, the rate of catalyst flow in a reformer is very slow. Whereas FCC catalyst circulation is measured in tons per hour, reforming catalyst circulation is a few hundred or a few thousand pounds per hour.

The moving-bed reformer operates in a semiregenerative mode by shutting down the regeneration section. In fact, a number of new units are designed to operate initially as semiregenerative, with the provision that a regeneration section can be added later.

The reactors in a moving-bed reformer are radial flow (see Fig. 3–3) and are either separated as in the semiregenerative unit or stacked one above the other.[1] The moving-bed unit operates at a lower pressure and a higher temperature (higher severity) than semire-

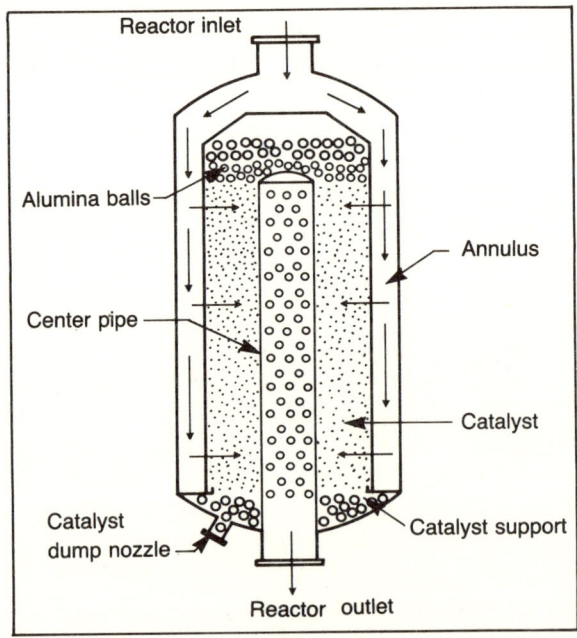

Fig. 3–3 Radial flow reactor

generative units because catalyst can be removed and regenerated before coke buildup seriously affects catalyst performance.

Current use of all of the above reformers is illustrated in the following data:[2]

Type of Unit	U.S. Capacity as of 1-1-84, bbl/sd
Semiregenerative	2,486,200
Cyclic	856,600
Moving-bed	400,000 (estimated)
Nonregenerative	120,000 (estimated)

The above are brief descriptions of reformer types. The selection of type depends on the specific needs of each refiner and deserves an in-depth study. Chapter 7 reviews the features of commercially licensed reforming processes.

REACTOR DESIGN

The reformer reactor must be designed to provide good flow distribution through the catalyst bed, that is, not allow total reactor feed to channel through or to bypass a portion of catalyst. The reactor should also be designed to allow a low pressure drop across the reactor. However, a minimum pressure differential is necessary for good flow distribution. Each licensor has its own reactor design, the details of which are proprietary.

In general, there are two types of reformer reactors: *radial flow* and *downflow.* In a radial flow configuration (Fig. 3–3), the total reactor feed enters the top of the reactor. A deflector baffle at the top of the reactor directs the vapor flow into the annular space between the reactor wall and the catalyst bed. Vapor flows radially through the catalyst bed toward the center and into a perforated pipe, called the center pipe. Vapor collected in the center pipe flows downward and out of the reactor.

The annular space—usually 4–12 in. across—is made possible by a screen arrangement that holds the catalyst away from the wall. The screens can be in the shape of scallops, as shown in Fig. 3–4. These scalloped screens are usually plates with slots small enough to prevent the catalyst from leaking into the annulus.

Some units use wire-mesh screen instead of scalloped plates (Fig. 3–5) for retaining the catalyst. Others use Johnson-type screens made of V-shaped wire welded spirally to a support.[3]

Scalloped screens are made in long sections small enough to be removed through the top inlet nozzle. These screens are supported at the bottom by an angle-iron shelf welded to the reactor shell.

36 CATALYTIC REFORMING

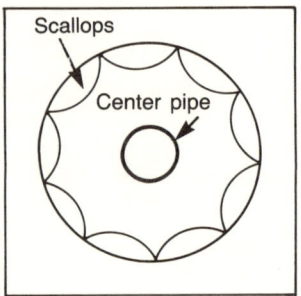

Fig. 3–4 Scalloped annulus

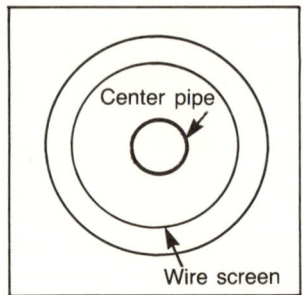

Fig. 3–5 Screen annulus

The center pipe, through which all vapors flow out of the reactor, is perforated in a manner providing proper pressure differential and flow distribution through the catalyst. The pressure differential across a freshly loaded radial-flow reactor should be in the range of 2–4 psi. The center-pipe holes are usually larger than the catalyst particles; so to prevent loss of catalyst, the center pipe is wrapped with a screen.

If an inert material such as firebrick, cement, sand, or alumina balls is used in the bottom of the reactor to support the catalyst, the holes in the center pipe are on the same level as the bottom of the catalyst bed. Some newer designs do not use inert materials in the bottom of the reactor, in which case catalyst fills the bottom head and center-pipe holes go all the way to the bottom.

In some units the center pipe is fitted into a socket at the bottom of the reactor and sealed with rope packing. The trend now seems to favor welding the center pipe in place.

One part of reactor internals that has been subject to numerous variations is the arrangement at the top, just above the catalyst bed. After a reactor comes on stream, some settling of the catalyst bed occurs. To avoid exposure of center-pipe holes and bypassing of catalyst, an excess quantity of catalyst called slump-and-seal catalyst is added. The slump-and-seal catalyst may be 6–18 in., depending on experience.

The deflector plate, which diverts inlet vapors to the annulus, has many variations. One arrangement describes a screen over the catalyst, with balls on the screen and a cover plate on the balls.[4] A cover plate only is shown for a moving-bed radial-flow reactor in a UOP patent.[5] In other units a few inches of space are left between the deflector plate and the catalyst side of the scallops. For these, support balls provide a seal against vapor bypassing the catalyst.

Radial-Flow Reactors

There are two types of radial-flow reactors in general use today: the *cold wall* and the *hot wall*. The reactor of Fig. 3–2 is the hot-wall type because insulation is outside the reactor shell and the inlet vapors are in contact with the inside wall of the reactor shell. This means reactor shell temperature is very close to that of the vapors (800–1,000°F).

The cold-wall reactor has insulation inside the reactor shell so that hot vapor does not contact the reactor shell; therefore, its temperature is lower (400–500°F). A cold-wall reactor with insulation inside the reactor shell covers the insulation with a metal liner overlay.[6] The purpose of the liner is to prevent leakage through cracks in the insulation, which would allow vapor to contact the metal shell. The serious consequences of reactor shell failure, a result of high temperature, are covered in chapter 5.

Downflow Reactors

Not all catalytic reformer reactors are radial flow.[7] A second type is the downflow reactor (Fig. 3–6). Vapor enters the top of this reactor,

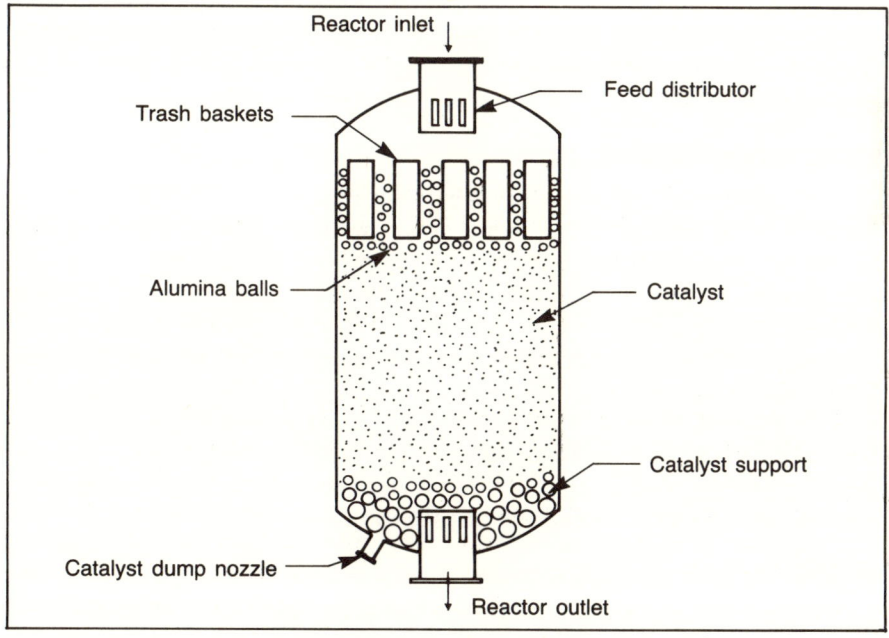

Fig. 3–6 Downflow reactor

38 CATALYTIC REFORMING

flows down through the catalyst, and flows out the bottom nozzle. Good flow distribution is provided by an inlet distributor and inert balls on top of the catalyst.

Another important feature is the *trash baskets* just above the catalyst. These are slotted or perforated pipes that are closed at the bottom end and wrapped with screen. They trap scale, catalyst fines, or other materials that plug voids in the catalyst bed and cause channeling of feed within the bed. The bottom of the catalyst bed is supported by a screen, support balls, or both.

This type reactor can also be used upflow with feed entering the bottom instead of the top. The reason this is not usually done is the reactor cannot be opened and catalyst skimmed off to alleviate high pressure drop due to crust formation or scale. With upflow, the entire catalyst bed must be dumped, screened, and replaced.

Some downflow reactors are spherical instead of cylindrical. In spherical reactors, the catalyst bed occupies little more than the bottom half of the sphere.

The pressure differential across a downflow reactor should be in the range of 5–10 psi when freshly loaded. Pressure drop will gradually increase with time on stream. If severe bed plugging occurs, the pressure drop may go as high as 40–60 psi. Operation then becomes limited, either by reactor metallurgy or by insufficient inlet pressure to maintain rate of flow.

RECYCLE COMPRESSOR

One other important equipment item that should be mentioned is the recycle compressor. The compressor must be able to circulate sufficient product separator gas to the combined-feed exchanger to maintain the desired hydrogen-to-hydrocarbon ratio at the inlet to the first reactor.

Recycle compressors are either reciprocating or centrifugal. The early reciprocating ones were equipped with carbon pistons for safety reasons, but teflon has replaced carbon. The later reformers are designed with a centrifugal compressor, either steam turbine or electrically driven. Because energy costs need to be minimized, centrifugal compressors can operate with variable-speed electrical motors.[8]

SUMMARY

Catalytic reformers have some things in common with other refinery units, but they also have requirements that necessitate unique design and operating features. A variety of possibilities in design and equipment are available for a reformer.[9] Anyone requesting of licensors the design, investment costs, and energy consumption for a new reformer soon realizes that many options are available.

REFERENCES

1. "Hydrocarbon Processing." In *1982 Refining Process Handbook,* 164, 166.
2. Cantrell, Ailleen, and L.R. Aalund. "Refining Survey." *OGJ* (26 March 1984):113.
3. NPRA 1982 Question-and-Answer Session (Q&A) on Refining and Petrochemical Technology, transcript p. 106, question 42.
4. Leg, Douglas J., and Ben G. Burke. "Method and Apparatus for Restraining Radial Flow Catalytic Reactor Center Pipes." U.S. pat. 4,374,095.
5. Greenwood, Arthur R., et al. "Multiple Stage Stacked Reactor System for Moving-bed Catalyst Particles." U.S. pat. 3,706,536.
6. Leg and Burke, op. cit.
7. Forbes, George. "Fixed-bed Catalytic Operations." U.S. pat. 3,431,084.
8. NPRA 1981 Q&A on Refining and Petrochemical Technology, transcript p. 103, question 33.
9. Hatch, W.H., S.J. Cohen, and R. Diener. "Modern Catalytic Reformer Designs Help Reduce Cost of Low-lead Gasoline." Paper AM–73–33 presented at the annual meeting of the NPRA, 1973.

CHAPTER 4

The Catalyst

The word "catalytic" in catalytic reforming means exactly that. The process is catalytic, and the heart of the process is the catalyst. The catalyst in the reactors promotes the desired reforming reactions and makes catalytic reforming a commercially feasible process.

Reforming catalysts are a marvel in their performance in commercial units and are getting better. They promote desired reactions and suppress undesirable reactions. These catalysts function efficiently with a wide range of feedstock composition. They can survive a variety of operating upsets and continue to reform in an acceptable manner. Finally, when their performance is no longer economically satisfactory, these catalysts can be regenerated, reactivated, rejuvenated, and restored to fresh-catalyst condition, ready for another operating cycle.

From the very outset of reforming development, the types of compounds most likely to be in the feedstocks and the types desired in the products were known. Most of the chemical reactions (see chapter 2) were recognized, although some of the intermediate reactions were, and still are, unknown. The chemists, engineers, and technicians assigned to catalyst development have made admirable progress since the early 1940s and have provided refiners with a remarkable selection of high-performance reforming catalysts (Table 4–1).

A fascinating review of the development of UOP catalyst, including the mistakes and frustration as well as the triumphs, is Vladimir Haensel's recollections in the Americal Chemical Society (ACS) symposium on selected histories of heterogeneous catalysis.[1]

EARLY DEVELOPMENT

To be commercially successful, a reforming catalyst (like all commercial catalysts) must possess three properties: activity, selectivity, and stability. These words are generously and loosely used in conversations and publications on catalysis but are rarely defined. The following illustrates these terms as they are used with respect to catalytic reforming and may or may not apply to other catalytic processes.

ACTIVITY

Activity generally means how well a catalyst does its job with respect to reaction rate, temperature, or space velocity. When considering a specific reaction, activity may be related to reaction rate. The higher the reaction rate, the higher the activity of the catalyst. For example, Gates identifies the activity of a copper-nickel alloy catalyst for the dehydrogenation of cyclohexane by plotting reaction rates vs atomic percent copper in the catalyst.[2]

Refinery engineers almost always refer to activity in a relative sense, i.e., one catalyst may be more or less active than another catalyst. Also, a catalyst may be less active after six months' operation than when it was fresh.

For motor fuel reforming, activity is generally represented by the temperature required to produce a given octane-number reformate. The lower the temperature, the more active the catalyst.

To illustrate, Fig. 4–1 represents two catalysts reforming the same feedstock at the same pressure and space velocity. As the run progresses,

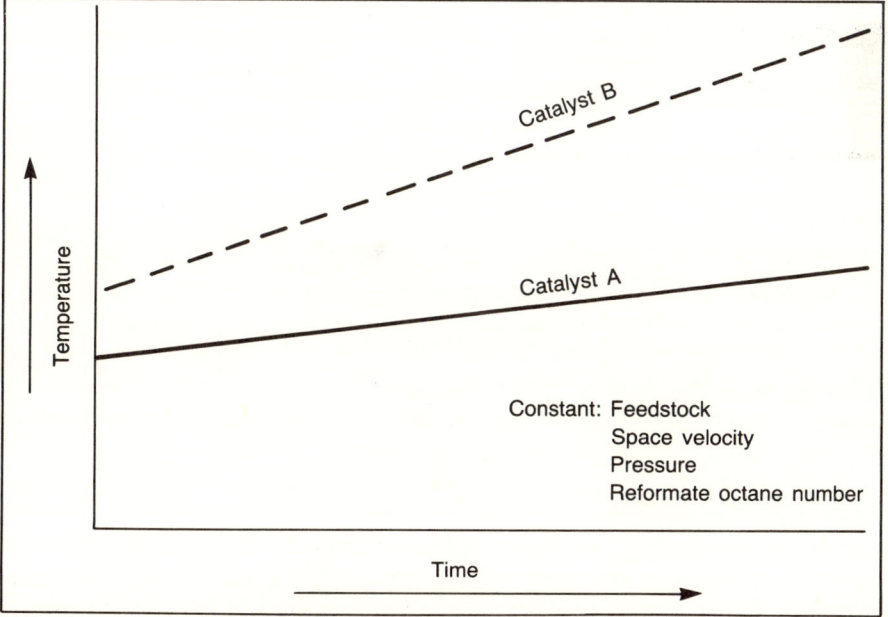

Fig. 4–1 Relative catalyst activity

TABLE 4-1
International reforming catalyst compilation[a]

Catalyst Designation	Primary Differentiating Characteristics	Application (Feedstock)	Application (Product)
CATALYTIC NAPHTHA REFORMING			
American Cyanamid Co.			
Aeroform© PHF–5 & PHF–5A	Monometallic; differing particle size	Naphtha	Gasoline or BTX
PRHF–30 PRHF–37 PRHF–50 PRHF–58	Bimetallic, but different amounts of active metals	Naphtha	Gasoline or BTX
KX–120 KX–130 KX–160	See Exxon Research & Engineering		
Trilobe© KX–130	See Exxon Research & Engineering Shaped extrudate		
Aeroform© PHF–4 & PHF–4A	Monometallic; different particle sizes.	Naphtha	Gasoline or BTX
Trilobe© P–8 & P–8A	Monometallic, high particle density, shaped catalyst; different particle sizes	Naphtha	Gasoline or BTX
Aeroform© PR–6 & PR–6A	Bimetallic; different particle sizes.	Naphtha	Gasoline or BTX
Trilobe© PR-8, PR-8A & PR-8E	Bimetallic, high particle density shaped catalyst; different particle sizes.	Naphtha	Gasoline or BTX
Chevron Research Co.			
Rheniforming Type F	Latest version, long cycle life stable product yields	Petroleum naphthas (SR and cracked) coal naphthas	Mogas and aromatics
Type E	1975 version, improved carrier	Petroleum naphthas (SR and cracked) coal naphthas	Mogas and aromatics

THE CATALYST 43

Form	Bulk Density (Compacted, lb/cu ft Unless Noted)	Carrier, Support	Active Agents	Availability to Refiner by Unrestricted Sale: √, Except Where Noted
Cylinder	~42	Alumina	P[b]	Licensees of Amoco Oil's ultraforming process
Cylinder	P	Alumina	P	Licensees of Amoco Oil's ultraforming process
				Licensees of Exxon's powerforming process
				Licensees of Exxon's powerforming process
Cylinder	~42	Alumina	Platinum/ chloride	Licensees of Exxon's powerforming process
Shaped extrudate	~44	Alumina	Platinum/ chloride	Licensees of Exxon's powerforming process
Cylinder	~42	Alumina	Platinum/ rhenium/ chloride	Licensees of Exxon's powerforming process
Shaped extrudate	~44	Alumina	Platinum/ rhenium/ chloride	Licensees of Exxon's powerforming process
Cylinder	40	P	Platinum/rhenium	Licensees Chevron rheniforming process
Cylinder	41	P	Platinum/rhenium	

TABLE 4–1
International reforming catalyst compilation[a]—cont'd

Catalyst Designation	Primary Differentiating Characteristics	Application (Feedstock)	Application (Product)
Cyanamid Ketjen			
CK–303 CK–304 CK–305 CK–306	All monometallics, but different active metal contents	Naphtha	Gasoline or BTX
CK–433	Bimetallic	Naphtha	Gasoline or BTX
CK–522 Trilobe©	Bimetallic, high particle density, shaped catalysts	Naphtha	Gasoline or BTX
PHRF–30	Multimetallic	Naphtha	Gasoline or BTX
CK–473	High-rhenium catalyst	Naphtha	Gasoline or BTX
KX–120 KX–130 KX–160	See Exxon Research & Engineering		
Engelhard Corp.			
E–301	Moderate stability	A–1 naphthas with feed sulfur <5 ppm wt	Motor fuel or aromatics
E–302	More stable than E–301	All naphthas with feed sulfur less than 10 ppm wt	Motor fuel or aromatics
E–311	Maximize LPG production	All naphthas with feed sulfur <5 ppm wt	LPG and motor fuel
E–601	High stability	All naphthas with feed sulfur <1 ppm wt	Motor fuel or aromatics
E–603	High stability. Lower metal loading than E–601	All naphthas with feed sulfur <1 ppm wt	Motor fuel or aromatics

Form	Bulk Density (Compacted, lb/cu ft Unless Noted)	Carrier, Support	Active Agents	Availability to Refiner by Unrestricted Sale: √, Except Where Noted
Cylinder	44	Alumina	Platinum/ chloride	Licensees Chevron rheniforming process
Cylinder	44	Alumina	Platinum/ rhenium/ chloride	Licensees Chevron rheniforming process
Shaped extrudates	45	Alumina	Platinum/ rhenium/ chloride	Licensees Chevron rheniforming process
Cylinder	P	Alumina	P	Licensees of Amoco Oil's ultraforming process
Cylinder	44	Alumina	Platinum/ rhenium/ chloride	
				Licensees of Exxon's powerforming process
Cylinder	45	Gamma alumina	Platinum	License/sale
Cylinder	45	Gamma alumina	Platinum	License/sale
Cylinder	45	Gamma alumina	Platinum	License/sale
Cylinder	45	Gamma alumina	Platinum & rhenium	License/sale
Cylinder	45	Gamma alumina	Platinum & rhenium	License/sale

TABLE 4-1
International Reforming Catalyst Compilation[a]—cont'd

Catalyst Designation	Primary Differentiating Characteristics	Application (Feedstock)	Application (Product)
E-611	Very high stability	All naphthas with feed sulfur <0.5 ppm wt	Motor fuel or aromatics

Note: Preceding catalysts available in higher density at 49.5 lb/cu ft. They are designated HD, for example E-301 HD.

Exxon Research & Engineering Co.

KX-120	Multimetallic catalyst for semiregenerative or cyclic units: different promoter	Virgin/cracked naphthas	High-octane motor gas & aromatics
KX-130	Different promoter	Virgin/cracked naphthas	High-octane motor gas & aromatics
KX-160	Different promoter	Virgin/cracked naphthas	High-octane motor gas & aromatics

Instituto Mexicano Del Petroleo

IMP-RNA-1	High activity bimetallic catalyst	Desulfurized naphtha	BTX aromatics, high-octane gasoline
IMP-RNA-2	Shape	Desulfurized naphtha	BTX aromatics, high-octane gasoline

Katalysatorewerke Huels AG

H 2440		Fraction 35–195°C	Gasoline, BTX
H 6413		Fraction 35–195°C	Gasoline, BTX
H 2462		Fraction 35–195°C	Gasoline, BTX

THE CATALYST 47

Form	Bulk Density (Compacted, lb/cu ft Unless Noted)	Carrier, Support	Active Agents	Availability to Refiner by Unrestricted Sale: √, Except Where Noted
Cylinder	45	Gamma alumina	Platinum & rhenium	License/sale
—	—	—	Platinum, etc.	Powerforming: process license through certified vendors
—	—	—	—	Powerforming: process license through certified vendors
—	—	—	—	Powerforming: process license through certified vendors
Cylinder	0.70 g/cc	Al_2O_3	Platinum/rhenium	Licensees
Trilobe© extrudates	0.70 g/cc	Al_2O_3	Platinum/rhenium	Licensees
Cylindrical extrudates	0.8 kg/l	Al_2O_3	Platinum 0.3–0.8% monometallic	Licensees
Cylindrical extrudates	0.6 kg/l	Al_2O_3		Licensees
Cylindrical extrudates	0.6 kg/l	Al_2O_3	Platinum 0.3-0.8% P	Licensees

48 CATALYTIC REFORMING

TABLE 4-1
International Reforming Catalyst Compilation[a]—cont'd

Catalyst Designation	Primary Differentiating Characteristics	Application (Feedstock)	Application (Product)
Procatalyse			
RG 402	H.P. reformer	Naphtha	Gasoline
RG 412	H.P. reformer	Naphtha	Gasoline
RG 422	M.P. reformer	Naphtha	Gasoline
RG 432	M.P. reformer	Naphtha	Gasoline
RG 482	M.P. & L.P. reformer	Naphtha	Gasoline
RG 442	Reformer	Naphtha	Gasoline & LPG
RG 451	Continuous regeneration reformer	Naphtha	Gasoline
CR 201	Continuous regeneration reformer	Naphtha	Gasoline
AR 401	Continuous regeneration reformer (Aromizing–IFP)	Naphtha	Aromatics
UOP Process Division			
R-8	Monometallic, oxidized	Unhydrotreated naphtha semi-regenerative	Gasoline, aromatics
R-9	Higher activity monometallic, oxidized	Semiregenerative or swing	Gasoline, aromatics

THE CATALYST 49

Form	Bulk Density (Compacted, lb/cu ft Unless Noted)	Carrier, Support	Active Agents	Availability to Refiner by Unrestricted Sale: √, Except Where Noted
Cylinder		Al_2O_3	Platinum 0.6%	Licensees
Cylinder		Al_2O_3	Platinum 0.35%	Licensees
Cylinder		Al_2O_3	Platinum 0.6% + Iridium	Licensees
Cylinder		Al_2O_3	Platinum 0.35% + Iridium	Licensees
Cylinder		Al_2O_3	Platinum 0.3% + Rhenium	Licensees
Cylinder		Al_2O_3	Platinum 0.35% + Iridium + promoter	Licensees
Sphere		Al_2O_3	Platinum 0.35% + Iridium + promoter	Licensees
Sphere		Al_2O_3	Platinum 0.35% + Sn	Licensees
Sphere		Al_2O_3	Platinum 0.6% + promoter	Licensees
Sphere	32	Alumina	Platinum	Licensees
Sphere	32	Alumina	Platinum	Licensees

TABLE 4-1
International reforming catalyst compilation[a]—cont'd

Catalyst Designation	Primary Differentiating Characteristics	Application (Feedstock)	Application (Product)
R-11	Higher activity monometallic, reduced and sulfided	Semiregenerative or swing	Gasoline, aromatics
R-12	Monometallic, higher stability than R-11, reduced and sulfided	Semiregenerative	Gasoline, aromatics
R-55	Monometallic, higher activity and stability than R-12, reduced and sulfided	Semiregenerative or swing	Gasoline, aromatics
R-15	Monometallic, reduced and sulfided	Semiregenerative	LPG, gasoline
R-16H	Bimetallic, higher stability than monometallics, reduced and sulfided	High severity, semiregenerative	Gasoline, aromatics
R-16F	Bimetallic, low platinum, higher stability than monometallics, reduced and sulfided	High severity, semiregenerative	Gasoline, aromatics
R16G	Bimetallic, higher activity and stability than R-16F and H, reduced and sulfided	High severity, semiregenerative	Gasoline, aromatics
R-18	Bimetallic, higher activity and stability than R-16 series, reduced and sulfided	High severity, semiregenerative	Gasoline, aromatics
R-22	Bimetallic, higher yields than platinum/rhenium reduced state	Continuous catalyst regeneration (CCR) or semiregenerative	Higher production of octane-barrels and aromatics
R-30	Bimetallic, higher yields and activity than platinum/rhenium, higher platinum than R-32, reduced state	CCR or swing	Higher production of octane-barrels and aromatics
R-32	Bimetallic, higher yields and activity than platinum/rhenium, reduced state	CCR	Higher production of octane-barrels and aromatics

Form	Bulk Density (Compacted, lb/cu ft Unless Noted)	Carrier, Support	Active Agents	Availability to Refiner by Unrestricted Sale: √, Except Where Noted
Sphere	32	Alumina	Platinum	Licensees
Sphere	32	Alumina	Platinum	Licensees
Extrudate	Dense loaded 52	Alumina	Platinum	Licensees
Sphere	32	Silica/Alumina	Platinum	Licensees
Sphere	32	Alumina	Platinum/rhenium	Licensees
Sphere	32	Alumina	Platinum/rhenium	Licensees
Sphere	32	Alumina	Platinum/rhenium	Licensees
Sphere	37	Alumina	Platinum/rhenium	Licensees
Sphere	32	Alumina	P	Licensees
Sphere	35	Alumina	P	Licensees
Sphere	35	Alumina	P	Licensees

TABLE 4–1
International reforming catalyst compilation[a]—cont'd

Catalyst Designation	Primary Differentiating Characteristics	Application (Feedstock)	Application (Product)
R–50	Bimetallic, higher activity and stability than R–16 series, reduced and sulfided	High severity, semiregenerative	Gasoline, aromatics
R–60	Bimetallic, higher stability than R–50, high activity, reduced and sulfided	High severity, semiregenerative	Gasoline, aromatics
R–62	Bimetallic, lower platinum than R–60, higher stability than R–50, high activity, reduced and sulfided	High severity, semiregenerative	Gasoline, aromatics

[a] Aalund, L.R. "Unique Survey Spotlights Complex Catalyst World." *OGJ* 8 October 1984:55.
[b] Proprietary

temperature is increased to maintain a constant octane number (RON0) of reformate. Catalyst A has a higher activity than catalyst B because A requires a lower temperature to produce the same octane-number reformate. Catalyst A also maintains a higher activity than catalyst B. Both catalysts deactivate as the run progresses. The temperature shown on the ordinate of Fig. 4–1 may be expressed in different units (as long as the temperature is consistent throughout the run) because *relative catalyst activity* is being measured.

Reforming pilot units are usually designed with isothermal reactors, and that temperature is used. Commercial reactors are adiabatic, so in plant practice either weighted average inlet temperature (WAIT) or weighted average bed temperature (WABT) is used (see chapter 5). Some refiners prefer to use space velocity instead of temperature to compare catalysts' activities. The higher the space velocity, the more active the catalyst. Either liquid hourly space velocity (LHSV) or weight hourly space velocity (WHSV) may be used (see chapter 5).

A comparison of catalyst activities using WHSV is illustrated in Fig. 4–2. In this test, the KX-130 catalyst was about three times as active as the other catalysts at the start of the run and showed a slight increase in relative activity as the run progressed.[3]

The above discussion refers to catalyst activity with respect to motor fuel reforming for octane number. In BTX reforming for aromatics, the same parameters of temperature or space velocity are used for comparing catalyst activity. However, in BTX reforming, instead of com-

Form	Bulk Density (Compacted, lb/cu ft Unless Noted)	Carrier, Support	Active Agents	Availability to Refiner by Unrestricted Sale: √, Except Where Noted
Extrudate	Dense loaded 52	Alumina	Platinum/rhenium	Licensees
Sphere	45	Alumina	Platinum/rhenium	Licensees
Sphere	45	Alumina	Platinum/rhenium	Licensees

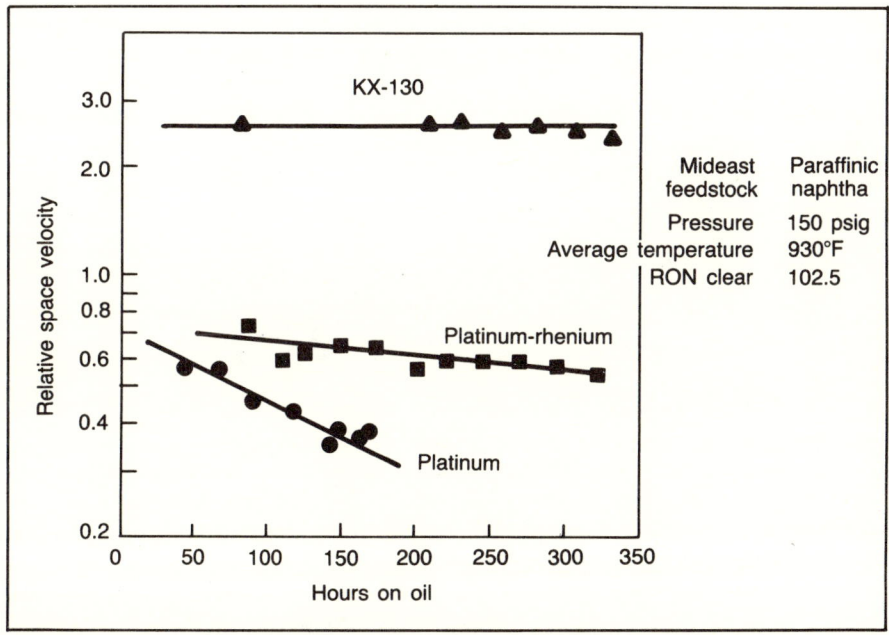

Fig. 4–2 Relative catalyst activity *(courtesy OGJ, after Cecil, Kmak, Sinfelt, and Chambers, ref. 3)*

54 CATALYTIC REFORMING

paring activity at constant octane number, constant conversion of a certain napthene such as methylcyclopentane (MCP) might be done. For convenience, some refiners still use octane-number response to reactor temperature to follow catalyst activity in a BTX unit.

To summarize, when a refiner thinks about activity of a reforming catalyst, it is almost always the temperature or space velocity one catalyst requires, compared to another catalyst, for the same octane number reformate.

In all this discussion about activity, nothing has been said about the percent yield from feed. That is *selectivity*. A catalyst can have good activity yet have poor selectivity, or vice versa. The best catalyst will have both good activity and good selectivity.

SELECTIVITY

In catalytic reforming, selectivity of the catalyst means the percent of desired product yielded from the feedstock. In motor fuel reforming, a high yield of reformate at the desired octane number is good selectivity. In BTX reforming, a high yield of total aromatics or of a desired specific aromatic is good selectivity.

For example, in Fig. 4–3, pilot unit data for reformate yield–octane number from two different catalysts are shown. Catalyst C has better

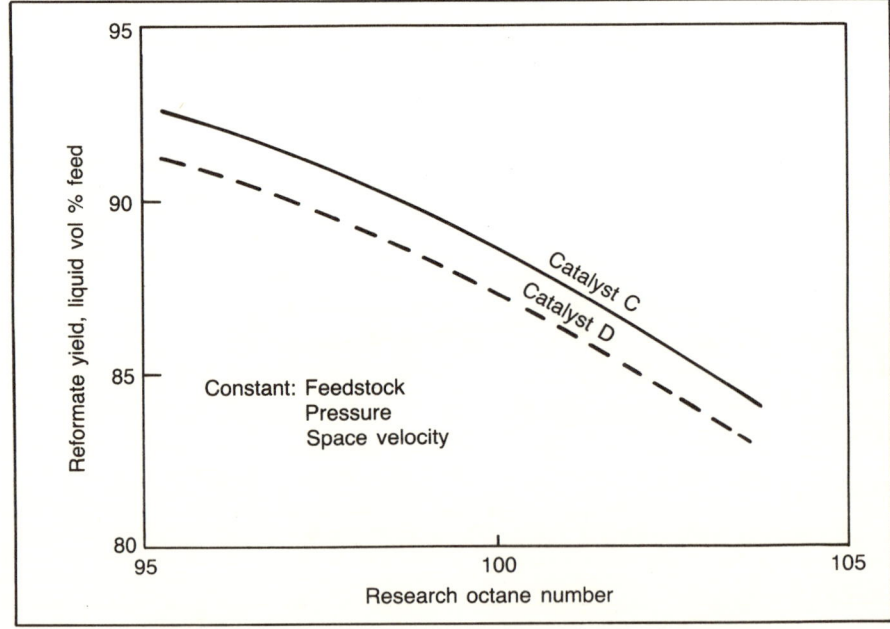

Fig. 4–3 Relative selectivity of reforming catalyst

selectivity than catalyst D because—on the same feedstock and at the same operating conditions (except reactor temperature)—catalyst C yielded 1–1.2 liquid vol % more reformate of the same octane number than did catalyst D. In these tests higher octane numbers were obtained by increasing reactor temperatures.

Table 4–2 illustrates selectivity in BTX reforming. The two platinum-containing catalysts have about the same activity for conversion of MCP, but the selectivity (yield of benzene) from MCP is far greater for the SiO_2 + Pt/SiO_2 catalyst than for the Pt/SiO_2 catalyst. Note also that the catalyst with no platinum has poor activity for MCP conversion compared to the other two.

TABLE 4–2
Conversion of methylcyclopentane catalyzed by acid, metal, and mixed catalysts[a]

Catalyst	Liquid Product Analysis, mol %			
	methylcyclopentane	methylcyclopentene	methylcyclopentadiene	benzene
10 cc SiO_2-Al_2O_3	98	0	0	0.1
10 cc Pt/SiO_2	62	20	18	0.8
SiO_2-Al_2O_3 + Pt/SiO_2	65	14	10	10.0

Reaction Conditions:
Temperature	500°C
Hydrogen partial pressure	0.8 atm
Methylcyclopentane partial pressure	0.2 atm
Residence time	2.5 sec
Catalyst	0.3 wt % Pt/SiO_2-Al_2O_3 420 sq m/g of surface area

[a]From Gates, et al. *Chemistry of Catalytic Processes.* McGraw-Hill, 1979.

Thus, when a refiner thinks about catalyst selectivity, he is looking at how much of a desired product his feedstock will yield. Usually, he will be comparing the selectivity of one catalyst with that of another.

STABILITY

A catalyst may have very high activity and excellent selectivity, but unless it also has stability it will be of no use to a refiner. Stability is the ability of a catalyst to maintain its activity and selectivity over a reasonable period. A catalyst that operates for a long time with little or no loss in activity or selectivity has good stability. A catalyst that quickly loses activity or selectivity has poor stability. A catalyst with good stability has a long cycle life between regenerations in a commercial unit.

A typical illustration of catalyst stability is Fig. 4–4. Obviously, catalyst B is far more stable and has a much longer cycle life than catalyst A with respect to both activity (reactor temperature) and selectivity (reformate yield). Fig. 4–4 also shows the significant advance in catalytic reforming with the development of bimetallic catalysts. It is not unusual for a bimetallic catalyst to have from two to four times the life of a platinum-only catalyst in the same service.

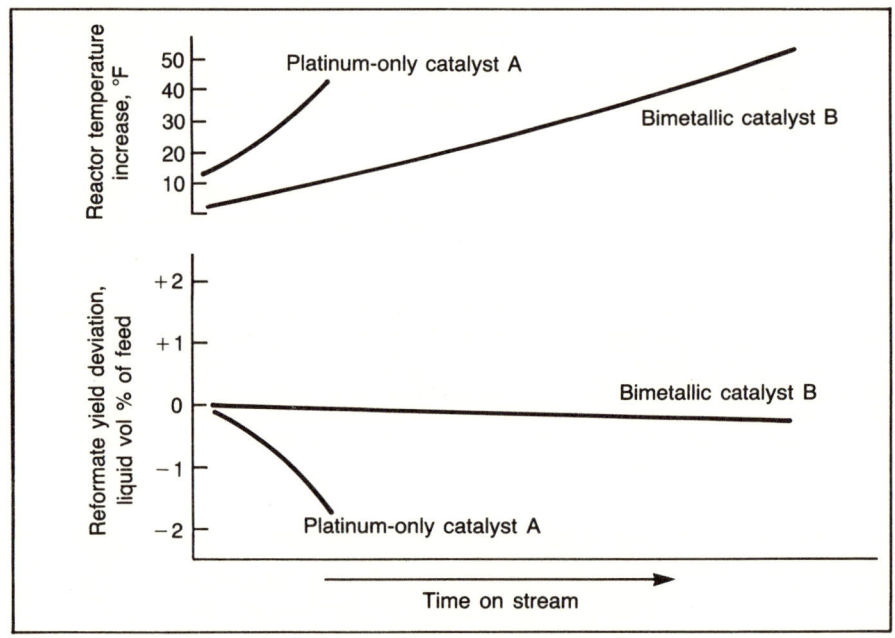

Fig. 4–4 Relative catalyst stability

Stability (catalyst life) is hard to predict accurately. Pilot units have such low charge rates that runs must extend for months to reach the end of a cycle. Commercial units vary too much in feed rate, feed composition, and unit interruptions to get stability measurement with a high degree of accuracy. Catalyst suppliers are reluctant to quote catalyst cycle-life estimates, particularly if the cycle life is more than one year or more than 100 bbl/lb. The best comparisons are relative stabilities obtained in pilot unit tests at very high severities to accelerate catalyst deactivation and complete a cycle in 20–30 days.

The foregoing discussion identified catalyst stability with activity and selectivity. A catalyst is sometimes referred to as being stable when it has

THE CATALYST 57

a higher tolerance for certain poisons such as sulfur. In another sense, a catalyst can have stability with respect to its ability to recover from an inadvertently large dose of poison such as sulfur.

Thus, when a refiner considers reforming catalyst stability, he thinks of how long his reformer can stay on stream before the catalyst must be regenerated or replaced. Except for extraordinary upsets in operation, there are generally two reasons for ending a cycle and shutting down to regenerate the catalyst.

The first and very common cause to end a cycle is loss of catalyst activity to the point that reactor inlet temperature has reached an upper limit and reformate octane number can no longer be maintained because temperatures cannot be raised. Reactor temperature may be limited by metallurgy of the reactors or by excessive hydrocracking of the feed.

The second reason for ending a cycle is either the catalyst has lost selectivity to the point that it is no longer economical to continue operation or hydrogen yield has become so low that the reformer can no longer supply hydrogen to other units or maintain its own hydrogen-to-hydrocarbon ratio.

A refiner is happiest when his reformer catalyst shows good activity, selectivity, and stability.

CATALYST DEVELOPMENT

Considering the catalysts used in reformer reactors today and that the first commercial reformer went on stream in November 1940,[4] the progress made in so few years is astonishing.

Reforming catalysts are heterogeneous catalysts. They are produced in a variety of physical forms and sizes. The predominant forms are $\frac{1}{16}$-, $\frac{1}{8}$-, and $\frac{1}{4}$-in. diameter extrudates (cylinders) or spheres. The lengths of these cylinders vary from $\frac{1}{4}$ to $\frac{3}{4}$ in. The extrudates are also available in the shape of a trilobe (cloverleaf) of about $\frac{1}{20}$-in. diameter.

The catalysts are composed of a base or support material, usually alumina (Al_2O_3), to which has been added certain metals catalytically active for the desired reforming reaction. The metals (examples: platinum, iridium, rhenium, germanium) comprise less than 1 wt % of the catalyst and are highly dispersed on the surface and in the pores of the alumina support.

The catalyst is manufactured in a manner that deposits the metals on the catalyst as extremely small, almost atomic-size crystallites. When reduced with hydrogen to the metallic state, these crystallites become active sites for catalytic reforming reactions. To make the catalyst even more active, a promoter such as chloride or fluoride is added to enhance the isomerization reactions.

58 CATALYTIC REFORMING

CATALYST PREPARATION

The preparation of the alumina support and the addition of metals and halide is very confidential. Two methods are commonly used.[5,6] In *impregnation,* the alumina base is first precipitated, washed, dried, formed, and then impregnated with metallic solutions. The final product is dried, and the halide promoter is incorporated. Sometimes the halide is added with the metals (for example, impregnation with chloroplatinic acid).

The second method of catalyst preparation, *coprecipitation,* involves mixing two or more solutions, followed by precipitation, washing, drying, forming, and drying. The impregnation method is preferred by manufacturers of reforming catalyst.

DEVELOPMENT OF REFORMING CATALYST

At this point, it would be well to review the development of reforming catalyst. The first catalysts were state-of-the-art type, using chromia oxides or molybdenum oxides to catalyze the hydrogenation and dehydrogenation reactions required to produce high-octane aromatics (chapter 2).

The support or base material was the subject of extensive research. Steiner demonstrated that molybdenum and chromia oxides were inactive by themselves but were active and stable when supported by alumina.[7] Silica and silica alumina were also used, but Haensel reports that, with silica alumina, hydrocracking was more difficult to control than with alumina alone.[8] Thus, before the introduction of platinum on alumina catalyst by UOP in 1949, reforming catalysts were predominantly 9–10% molybdenum oxide on alumina, in the shape of granules or pellets of 2–4 mesh, equivalent to $5/16$- to $3/16$-in. diameter.[9,10]

The development of reforming catalyst followed what is sometimes called trial and error or, as Schuitt and Gates put it, "Edisonian" means. They add that our understanding of these catalysts was still "primitive" in 1983. The point is well made by Haensel that experimental catalysts must be properly tested and the results correctly interpreted for real progress. Looking back over forty-plus years of catalyst development, the progress seems to have come in giant steps:

I   Better-quality alumina base gave longer life (stability)
II Platinum replaced molybdenum and chromia, giving better yields
III Regeneration technology gave redispersion of platinum, restoring catalyst activity and selectivity to that of fresh catalyst
IV Bimetallic and multimetallic catalysts markedly improved activity, selectivity, and stability

Since the introduction of bimetallics in 1967, improvements have been slow and mostly in the structure of the alumina base. However,

THE CATALYST

there is no doubt that the catalysts offered in 1985 are better than the 1967 bimetallics.

MULTIFUNCTIONAL CATALYST

An outstanding feature of reforming catalyst, even from the beginning, is the incorporation of more than one function in a single catalyst. For this reason, reforming catalysts have been termed dual functional, bifunctional and, recently, multifunctional.[11,12,13,14]

Technologists involved in the development of reforming catalysts recognized that the reforming reactions of chapter 2 required at least two different functions. One was a metal to catalyze the dehydrogenation of naphthenes to aromatics and to hydrogenate olefins formed by dehydrogenation and hydrogenolysis of paraffins. It is generally acknowledged that the metal component also contributes to dehydrocyclization and isomerization. The second function needed was an acid to catalyze isomerization, cyclization, and hydrocracking. The contribution of catalyst components to reactions that occur in reforming and the effect of environmental variables, although the subjects of vast research, are still not well understood. The general consensus is that the catalyst-metal function and the acid function are both necessary and both contribute, sometimes individually and sometimes collectively.

One of the better visualizations of the dual function of reforming catalyst and reforming reactions is Fig. 4–5. This chart appeared in a

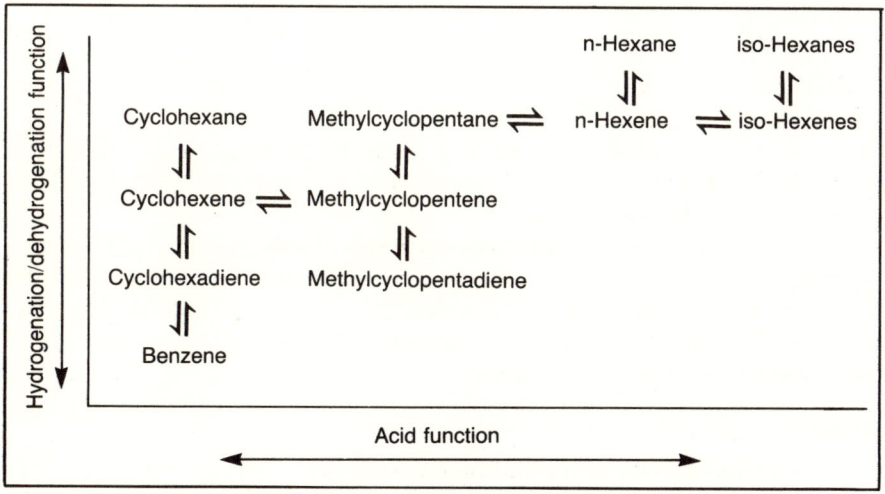

Fig. 4–5 Reactions of a multifunction reforming catalyst *(courtesy Industrial and Engineering Chemistry, © American Chemical Society, 1953, after Heinemann, Mills, Hattman, and Kirsch, ref. 15)*

number of reforming articles and was first published by Mills et al.[15] In this simplified representation of reforming reaction network, reactions indicated by vertical arrows are catalyzed by the metal function of the catalyst. Reactions indicated by horizontal arrows are catalyzed by the acid function of the catalyst. For example, the mechanism for producing benzene from n-hexane is dehydrogenation to n-hexene on a metal site, isomerization of n-hexene to methylcyclopentane on an acid site, dehydrogenation to methylcyclopentene on a metal site, isomerization to cyclohexene on an acid site, then dehydrogenation on a metal site to cyclohexadiene, and, finally, benzene.

CATALYST COMPOSITION

The metal dehydrogenation-hydrogenation component of a catalyst in commercial reforming reactors in 1985 must surely be platinum. A comparison of the dehydrogenation activity of various metals on oxide supports reported by Ciappeta shows platinum alumina to be 500–1,000 times more active than chromia alumina or molybdenum alumina.[16] None of the other metals—not even iridium or palladium—approaches the activity of platinum on alumina. In bimetallic catalysts, the role of the second metal (such as rhenium, iridium, or germanium) is not definitely known except that in some way it attenuates the effect of coke on catalyst. Some claim the second metal also increases activity or selectivity.

The acid function of commercial reforming catalysts is provided by the support or base, which in 1985 must certainly be alumina, promoted by a halogen such as chloride. Some early catalysts used platinum on silica alumina, which has a very strong acidity (so strong that hydrocracking was difficult to control). Gates points out that alumina itself is too weakly acidic to catalyze skeletal isomerization.[17] Therefore, for catalytic reforming reactions requiring stronger acidic sites, the acidity is induced by treatment of alumina with halogens such as hydrogen chloride or hydrogen fluoride.

Some fluorided alumina catalysts are still commercially available; but because acidity can be more easily controlled with chloride, catalyst supports are now predominantly chlorided alumina. Alumina does not tightly hold chloride. Control and adjustment of chloride in commercial reforming is discussed in chapter 5.

Aluminas

The support or base for the metal component of reforming catalyst is alumina. In fact, since the metal content of reforming catalyst is rarely more than 1 wt % and chloride is about 1 wt %, reforming catalyst is approximately 98% alumina.

Alumina exists in a number of forms (called alumina hydrates) which may be transformed by heating (see Fig. 4–6).

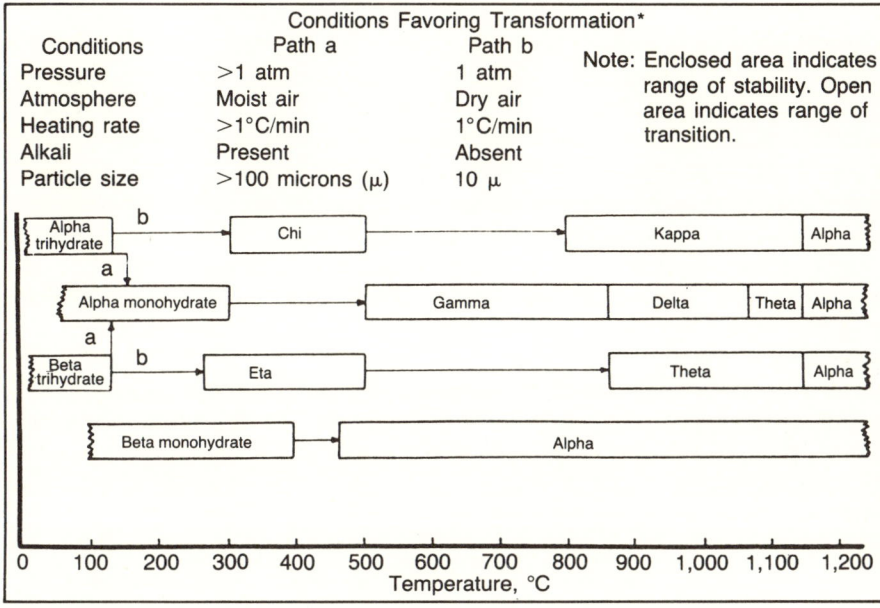

*after Pollak

Fig. 4–6 Decomposition sequence of alumina hydrates *(courtesy Alcoa)*

Alumina support for reforming catalyst is either the gamma or the eta form. According to Edgar, the eta form has a higher acid function than the gamma and a higher initial surface area but loses surface area after a few process cycles and regenerations.[18] The gamma form is more stable thermally than the eta. Thus, the gamma form is probably the form used in most reforming catalysts expected to go through a number of regenerations. When a temperature excursion has possibly occurred in a reforming reactor, the detection of alpha alumina in the catalyst will confirm the catalyst has been subjected to 1,900°F or higher.

COMMERCIAL CATALYSTS

The preparation and exact composition of catalyst are among the most confidential and highly classified aspects of commercial reforming catalysts. Although a catalyst supplier provides instruction on use of a proprietary catalyst, very little about catalyst properties is revealed, even under confidentiality agreements. The best a nonconfidential publication can do is to survey information disclosed in patents on reforming catalysts.

A great number of U.S. patents have been issued on the preparation, composition, and use of reforming catalysts. As with most patents, an effort is made to make the claims as broad as possible yet have a strong, valid patent. For example, a patent may claim the platinum content of a reforming catalyst ranges 0.1–2.0% by weight. In actual practice, however, a catalyst supplied commercially will have a platinum content controlled within narrowly specified limits, say, 0.30–0.31% by weight.[19]

Two other disclosures within a patent provide a better clue concerning the concentration of a catalyst component as to which catalyst will perform best. The first is a disclosure of the "preferred" concentration. The second is examples of catalysts whose specific compositions are disclosed.

For example, the above patent states that the platinum concentration generally ranges from 0.1–2 wt % but that the preferred concentration is from about 0.2–0.6 wt %. For best performance, the catalyst should not contain less than 0.2 wt % nor more than 0.6 wt % platinum. Further, the patent discloses examples of five multimetallic catalysts. In each case, the platinum content is 0.3 wt %. In addition to platinum, the catalysts contain iridium, selenium, copper, and chloride on gamma alumina.

The compositions of reforming catalysts selected from U.S. patents are listed in Table 4–3. There is no certainty that the compositions shown are exactly that of any commercial catalyst. But in the author's experience, these are in the range of commercial reforming catalysts. A component of nearly all reforming catalysts (not shown in Table 4–3) is sulfur. Most patents (for example, U.S. patents 4,210,524; 4,312,778; 4,379,076) teach that freshly reduced catalyst must be sulfided prior to reforming. The sulfiding may be done by the catalyst supplier or by the refiner as part of the unit start-up procedure (chapter 5).

The amount of sulfur on the catalyst is in the range of 0.1–0.5 wt %. The purpose of sulfiding is to attenuate or, as sometimes stated, to take the edge off of superactive sites always present on fresh catalyst. Otherwise, excessive hydrocracking may occur to the extent that the catalyst is degraded and cycle life is reduced. Sulfiding must be done with either new catalyst or regenerated catalyst.

After start-up, most of the sulfur will be stripped from the catalyst as hydrogen sulfide and removed from the system in the product separator gas and stabilizer overhead gas. The superactive sites on the catalyst by then will have been rendered inactive by coke or other feed contaminants. It may seem contradictory that sulfur, which is a tem-

TABLE 4-3
Composition of reforming catalysts (as cited in U.S. patents)

Catalyst	Monometallic[a]		Bimetallic[b]		Bimetallic[c]	
Component	Preferred Range	Specific Example	Preferred Range	Specific Example	Preferred Range	Specific Example
Platinum, wt %	0.01–1	0.3	0.2–1	0.7	0.05–1	0.375
Rhenium, wt %	—	—	0.1–2	0.7	0.05–1	0.375
Chloride, wt %	—	0.45	0.1–3	Not shown	0.5–1.5	1.0
Fluoride, wt %	0.1–3	0.3	—	—	—	—

Catalyst	Multimetallic[d]		Multimetallic[e]		Multimetallic[f]	
Component	Preferred Range	Specific Example	Preferred Range	Specific Example	Preferred Range	Specific Example
Platinum, wt %	0.05–1	0.375	0.01–1.2	0.3	0.2–0.6	0.3
Rhenium, wt %	0.05–1	0.375	0.01–0.2	0.03	—	—
Germanium, wt %	0.01	0.05	—	—	—	—
Iridium, wt %	—	—	0.01–0.1	0.025	0.2–0.6	0.3
Copper, wt %	—	—	—	—	0.025–0.08	0.05
Selenium, wt %	—	—	—	—	0.01–1	0.04
Halide (Chloride or Fluoride), wt %	0.5–1.5	1.0	0.1–2	Not shown	0.7–1.2	0.9

[a] U.S. pat. 2,752,289
[b] U.S. pat. 3,415,737
[c] U.S. pat. 4,210,524
[d] U.S. pat. 4,312,788
[e] U.S. pat. 3,487,009
[f] U.S. pat. 4,379,076

64 CATALYTIC REFORMING

porary poison, is deliberately added to the catalyst; but experience shows this is the best way to control fresh-catalyst activity.

Metal Dispersion

Let us now discuss the dispersion of metals on the catalyst base. Platinum, rhenium, and other metals exist on the alumina substrate as crystallites. A great deal of effort has been expended to accurately determine the size of platinum and rhenium crystallites on reforming catalyst.[20,21] The data indicate the size of these platinum particles is in the range of 8–100 angstroms (Å). A fresh highly active catalyst shows metal crystallite size as less than 35 Å (preferably less than 24 Å). Gates shows that a platinum crystal with regular faces, containing a total of six atoms, has a crystal edge length of 2 atoms or 5.50 Å. If the length of the crystal edge is 10 atoms or 27.50 Å, there will be 670 total atoms in the crystal. These figures give some idea of the minute size of metal crystallites on platinum catalyst. The general conclusions of studies on metal crystallites are that, the smaller the crystallite size, the more active and more selective the reforming catalyst.

Catalyst Density

Catalyst properties other than composition are important to plant engineers. One property of practical concern is catalyst bulk density, called average bulk density (ABD) or loading density in pounds per cubic foot. The importance of ABD is that reforming unit reactors have a maximum catalyst volume capacity and that almost all reforming catalyst is priced and sold by weight. Therefore, an engineer needs the ABD to determine how many pounds of catalyst are required to load the reactors, how much the total cost will be, and how much metal (platinum, rhenium, iridium, etc.) must be made available to the manufacturer. The following illustrates such a calculation.

Reformer charge 20,000 barrels per stream day (b/sd) (833.33 bbl/hr)
Space velocity, LHSV 1.0 bbl/hr (bbl of catalyst)
Catalyst ABD 40 lb/cu ft
Platinum content 0.30 wt %
Vol of catalyst = 1.0 × 833.33 bbl or
833.33 × 5.61 cu ft/bbl = 4,765 cu ft
Wt of catalyst = 4,765 × 40 = 190,600 lb
Platinum required: $190{,}600 \times 0.003 \times 14.583 \frac{\text{troy oz}}{\text{lb (avoir)}} = 8{,}338.6$ troy oz

The engineer now knows that 190,600 lb of catalyst and 8,338.6 troy oz of platinum are required. These are estimates because the measurement of catalyst as loaded and platinum assay (analysis for platinum content) of the catalyst as delivered are the best numbers and may differ from the calculated ones. The ABD of reforming catalysts ranges from 30 to 55 lb/cu ft. Each catalyst supplier can provide a good estimate based on experience in loading commercial reactors.

Platinum and Rhenium

The platinum is by far the most costly part of the catalyst. Fig. 4–7 shows the dealers' price of platinum over a six-year period. At the peak dealers' platinum price, the 8,338.6 troy oz in the above illustration would cost over $8 million, which is equivalent to about $41/lb catalyst.

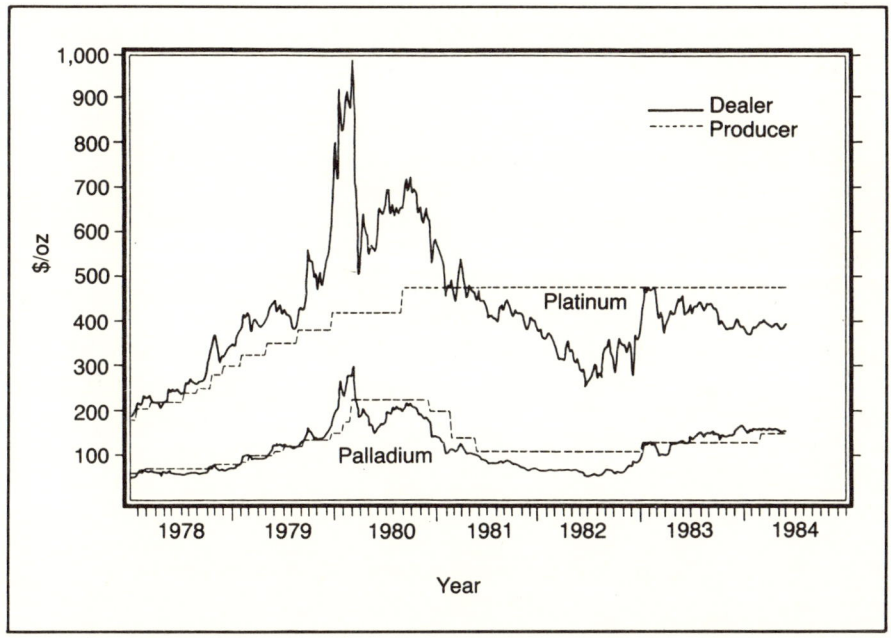

Fig. 4–7 Platinum and palladium prices *(courtesy J. Aron & Co.)*

In the early 1980s, with platinum ranging from $300–$400/troy oz, the cost of the platinum on the catalyst was equivalent to $13–17.50/lb catalyst. Because of the fluctuation in platinum prices, users of reformer catalyst commonly own their platinum, which is maintained in a platinum account or "bank" with catalyst suppliers or dealers in precious metals.

Spent catalyst (including fines) is returned for metals recovery. The refiner is credited for 99–99.5% of the platinum assayed on the spent catalyst.

Rhenium is generally priced by the pound. When platinum/rhenium bimetallic catalysts were first introduced in 1967, rhenium was in short supply and as high as $3,000/lb. It has since become plentiful and sold for about $300/lb or less in the early 1980s.

Recovery of rhenium from spent catalyst is more difficult than recovery of platinum. A return of only 85–95% of assayed rhenium is normal. Because of its low price and low recovery, some refiners do not request that rhenium be recovered from spent catalyst.

Surface Area

Surface area of a new catalyst generally is not a user specification and may or may not be provided by the catalyst manufacturer on request. Surface area to a refiner indicates catalyst degradation through a number of regenerations. Some catalysts' surface areas decrease on each regeneration. Over time, the surface area loss can decrease catalyst activity, selectivity, and cycle life.

Patent literature discloses surface area of alumina base before impregnation and calcining is in the preferred range of 100–500 sq m/g.[22,23,24] In the author's experience, the surface area of finished reforming catalyst ranges from 150–250 sq m/g. The change in surface area over time for a bimetallic catalyst in commercial service is shown in Table 4–4. After 7½ years and four regenerations, the catalyst lost 15–21% surface area; loss of catalyst selectivity, activity, or stability was not observed in unit operations. At the 1982 NPRA refining Q&A, D.M. Torchia reported that a catalyst from a cyclic unit with over 700 regenerations had a surface area just a little under 100 sq m/g.[25]

TABLE 4–4
Surface area change in a bimetallic catalyst in commercial service

	Surface Area of Bimetallic Catalyst,[a] sq m/gm			
	New Catalyst[b]	After 2.5 Years 1 Regeneration[c]	After 6 Years 3 Regenerations[c]	After 7.5 Years 4 Regenerations[c]
Reactor 1	185	181	154	157
Reactor 2	185	179	155	154
Reactor 3	185	180	155	156
Reactor 4	185	176	156	146

[a]After laboratory regeneration (except new catalyst)
[b]Surface area given by catalyst supplier
[c]Plant laboratory tests

Catalyst Size

The size and shape of reforming catalysts are matters of preference of the licensor or manufacturer. Commercial reforming catalysts are commonly $1/16$-, $1/8$-, or $1/4$-in. diameter and are shaped as spheres or cylindrical extrudates. Some extrudates are the shape of a three- or four-lobe cloverleaf. The smaller sizes have higher activities and generally comprise the bulk of reactor fill. The larger diameters are used at the bottom of the bed as catalyst support, at the top of the bed for flow distribution, and as a scale and fines trap.

Dense loading or *catalyst-oriented packing* (COP) techniques are utilized more and more for reforming catalyst. These techniques load 5–7 wt % more spheres and 10–15 wt % more extrudates than the sock-loading method.

The advantage of lower pressure to increase yield of reformate is discussed in chapter 5. Catalyst size significantly affects pressure drop across a bed of catalyst. Fig. 4–8 demonstrates that, in laboratory tests, $1/16$-in. diameter cylinders doubled the pressure drop per meter of bed depth, compared to $1/8$-in. diameter cylinders. However, most commercial reforming units are designed to use $1/16$-in. diameter catalyst.

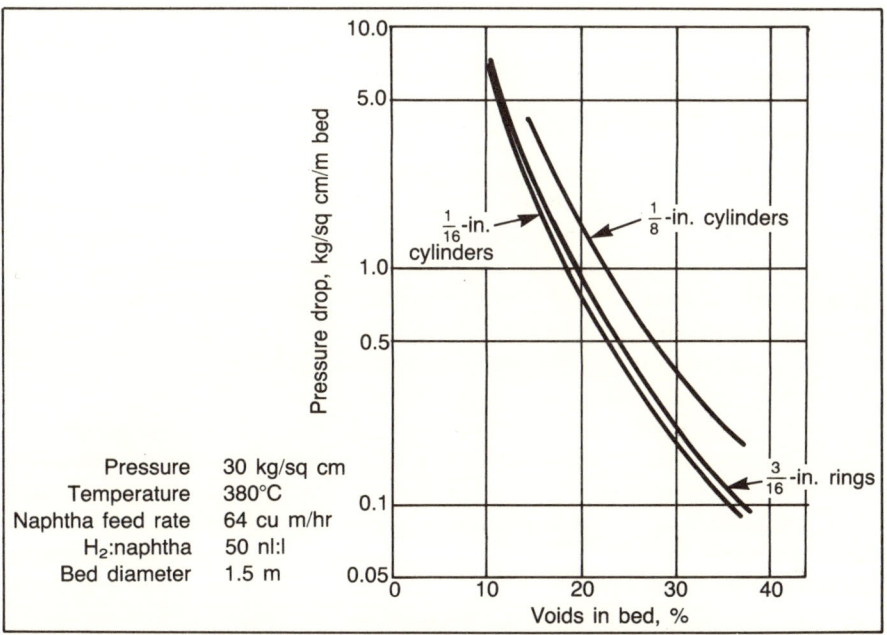

Fig. 4–8 Effect of catalyst size on pressure drop *(courtesy Haldor-Topsoe)*

Crush Strength

Crush strength of reforming catalysts seldom presents a problem. A minimum single-particle crush strength of 4 lb (by ASTM method D–4179–92) is satisfactory. Crush strengths of 10 lb or higher are the general rule for reforming catalyst extrudates and are much higher for spheres. A bulk crush strength method for particles is being developed by ASTM. When finalized, this method should be more applicable to catalyst beds than the single-particle method.

CATALYST POISONS

There are certain substances which 40 years' experience indicates are most likely to appear in reforming feedstocks and which act as poisons to the catalyst. These are called poisons because, in one way or another, they impair the proper performance of the catalyst. In commercial operations, the poisons usually enter the reformer as feed contaminants but occasionally enter through water or chloride injection streams.

Reforming catalyst poisons are classified as either temporary or permanent. *Temporary poisons* are those which can be removed from the catalyst without a shutdown, and the catalyst is restored to proper activity and selectivity. A classic temporary poison is sulfur which, when present in amounts greater than 0.5 ppm wt, causes a significant activity loss (octane number decrease) with a bimetallic catalyst. However, if the sulfur content of the feedstock is brought down to 0.5 ppm wt, the catalyst will slowly recover and activity will be restored to the level of pre-high-sulfur feed.

A *permanent catalyst poison* is one that impairs catalyst performance to the point that the unit must be shut down and the catalyst replaced because there is no way to restore it to a satisfactory performance level. A classic permanent catalyst poison is arsenic, which deposits on the catalyst and causes a rapid loss in activity. A general requirement is that the maximum allowable arsenic content of the feedstock is 1 part per billion (ppb) wt, which is as low as analytical methods can detect arsenic in naphtha feedstocks.

One other substance that markedly affects reforming catalyst but which is neither a temporary nor a permanent poison by the above definitions is coke. Coke (or carbon) accumulates on reforming catalyst and gradually builds up until the unit must be shut down. In the case of a cyclic unit, a reactor is taken off stream and the catalyst is regenerated by burning off the coke. After the carbon burn, the catalyst is restored to fresh activity and selectivity by a rejuvenation procedure.

Following are the poisons most often encountered in commercial reforming:

Temporary Poisons	Approximate Maximum Allowable in Reformer Feed[a] ppm wt
Sulfur	0.5–1
Nitrogen	0.5–1
Chlorides	0.5
Fluorides	0.5
Water or oxygenated hydrocarbons	[b]2–4
Permanent Poisons[c]	**ppb wt**
Arsenic	1-2
Lead	10-20
Copper	Not available[d]
Silica	Not available[d]
Phosphorus	Not available[d]

[a]For bimetallic catalyst. Limits are for illustrative purpose and may be higher or lower, depending on specific catalyst.
[b]Not very precise figures. Analysis is difficult. Should be low enough to give 5–10 mol ppm water in recycle hydrogen without water injection.
[c]Lists only those most commonly known. Any metal should be suspect.[26]
[d]Similar to arsenic and lead.

The effect of poisons on catalyst and catalyst performance in a commercial reformer, how the presence of poisons may be recognized, and steps to eliminate poisons are discussed in chapter 5. A brief discussion of the interaction of a poison and a reforming catalyst is given here.

Sulfur. Sulfur deactivates the hydrogenation/dehydrogenation function of the catalyst by forming metal sulfides. However, the reaction is reversible. When sulfur is removed from the feed, the sulfur leaves the catalyst and appears as hydrogen sulfide in reformer products.

Nitrogen. Organic nitrogen compounds in reformer feed are converted to ammonia, which poisons the acid function of the catalyst. This generally means the ammonia reacts with the chloride component. When organic nitrogen is eliminated from the feed, the catalyst gives up ammonium chloride and recovers its acid function if chloride is brought up to the proper level by chloride addition. Molecular nitrogen (N_2) apparently does not react at reforming conditions and so is not a catalyst poison.

Halides. Too much chloride increases the acid function of the catalyst beyond the optimum required for balanced catalyst performance. When excess chloride is eliminated from the feed, chloride leaves the catalyst. Balance is restored by control of chloride injection. Fluoride has the same effect as chloride, except fluoride is very difficult to strip from the catalyst and makes controlling the acid-catalyzed hydrocracking reaction more difficult.

Water. Water comes from moisture in the feed or from oxygenated hydrocarbons converted to water at reforming conditions. Although some moisture is required to activate the alumina support, excess water can throw the acid function of the catalyst out of balance by stripping chlorides from the catalyst. Balance of catalyst functions is restored by elimination of the excess water source and by chloride injection.

Metals. Trace quantities of metals such as arsenic or lead can enter a reformer with reactor charge. The metals deposit nearly quantitatively on the catalyst and deactivate the metal (platinum) function of the catalyst. The effect is irreversible; the arsenic or lead cannot be removed to restore the reforming functions of the catalyst. Preventing arsenic or lead from entering the unit and replacing the poisoned catalyst are necessary to bring unit performance up to its proper level.

Carbon on catalyst. Rapid buildup of carbon on the catalyst accelerates loss of activity and selectivity. Carbon laydown is generally attributed to condensation of aromatics into polycyclic aromatics. These polycyclic aromatics strongly adhere to the catalyst and are ultimately converted to coke. At some level of coke on the catalyst—about 7–10 wt % for platinum-only catalyst and 20–40 wt % for bimetallic catalyst—activity and selectivity are so low that the catalyst must be regenerated by burning off the coke. A high rate of carbon laydown and coking is frequently a result of high-boiling hydrocarbons in reactor charge.

As feedstock end point (ASTM distillation D86) increases above 400°F, increasing amounts of naturally occurring polycyclic aromatics are likely to be included in the naphtha. Therefore, a reactor charge limitation of 400°F end point is frequently specified. However, some refiners, willing to take short cycles and frequent regenerations, allow feed end points as high as 425–430°F. Others take short cycles and frequent regenerations by operating at high severity to obtain higher-octane-number reformate, which also accelerates carbon laydown on the catalyst.

SUMMARY

Reforming catalysts are a complex aggregation of catalytically active materials, designed to promote specific reactions. They have been the subject of much research and development and are still being intensively studied.[27,28] The foregoing is primarily directed to the characteristics of reforming catalysts. Their application and use in commercial units are covered in chapter 5.

REFERENCES

1. Davis, Burton H., and William P. Hettinger, editors. "Heterogeneous Catalysis, Selected American Histories." Paper presented at the 183d

meeting of the American Chemical Society (ACS), Las Vegas, 28 March–2 April 1982.

2. Gates, Bruce C., James R. Katzer, and G.C.A. Schuitt. *Chemistry of Catalytic Processes.* McGraw-Hill.

3. Cecil, R.R., W.S. Kmak, J.H. Sinfelt, and L.W. Chambers. "New Reforming Catalyst Highly Active." *OGJ* (17 August 1972): 50.

4. Davidson, Robert L. "Catalytic Reforming—How It Works—and the Processes You Can Use." *Petroleum Processing* (August 1955): 1174.

5. Buss, Walden C. "Reforming Process with Promoted Low-platinum-content Catalyst." U.S. pat. 3,578,583.

6. Higginson, George W. "Making Catalysts—An Overview." *Chemical Engineering* (30 September 1974): 98–104.

7. Steiner, H. *Discussions of Faraday Society* 8 (1950): 264–270.

8. Haensel, Vladimir. "The Platforming Process—Personal Recollections." Paper presented at the 183d meeting of the ACS, Las Vegas, 28 March–2 April 1983.

9. Davidson, op. cit.

10. Sittig, Marshall, and Wayne Warren. "How to Get Those Top Octanes." *Petroleum Refiner* 9(34) (September 1955): 230–280.

11. Cecil, et al., op. cit.

12. Schuitt, G.C.A., and B.C. Gates. "The Winning Catalysts are Multifunctional." *Chemtech* (September 1983): 556.

13. Ciapetta, Frank G. "Catalytic Reforming." *Petro/Chem Engineer* (May 1969): C-19–C-31.

14. Ciapetta, F.G., and D.N. Wallace. "Catalytic Naphtha Reforming." *Catalysis Reviews* 5(1) (1971): 67–158.

15. Heinemann, H., G.A. Mills, J.B. Hattman, and F.W. Kirsch. "Houdriforming Reactions." *Industrial and Engineering Chemistry (I&EC)* 1(45) (January 1953): 130–137.

16. Ciapetta, "Catalytic Reforming," op. cit.

17. Gates, et al., *Chemistry of Catalytic Processes,* op. cit.

18. Edgar, M. Dean. "Catalytic Reforming of Naphtha in Petroleum Refineries." *Applied Industrial Catalysis* 123(1) (1983).

19. Eberly Jr., Paul E., et al. "Reforming with Multimetallic Catalysts." U.S. pat. 4,379,076. 5 April 1983.

20. Gates, et al., *Chemistry of Catalytic Processes,* op. cit.

21. Ciapetta and Wallace, "Catalytic Naphtha Reforming," op. cit.

22. Eberly, et al., op. cit.

23. Mahoney, John A., and Albert Hensley Jr. "Reforming with Platinum-Rhenium-Selenium Catalysts." U.S. pat. 3,884,799. 20 May 1975.

24. Antos, George J. "Hydrocarbon Conversion with a Superactive Multimetallic Catalytic Composite." U.S. pat. 4,149,962. 17 April 1979.
25. NPRA 1982 Q&A on Refining and Petrochemical Technology, p. 104, question 36.
26. NPRA 1978 Q&A on Refining, p. 82, question 26.
27. Schuitt and Gates, "The Winning Catalysts Are Multifunctional," op. cit.
28. Charcosset, H. "Current Status of Basic Research on Industrial-type Bimetallic Platinum-based Catalysts on Alumina Supports." *International Chemical Engineering* 2(23) (April 1983): 187–212 and 3(23) (July 1983): 411–426.

CHAPTER 5

Process Variables and Unit Operation

Catalytic reformers are designed for flexibility in operation, whether for motor fuel or for aromatics (BTX) production. This chapter is devoted to motor fuel reforming. Chapter 6 covers BTX operation.

In the production of gasoline (chapter 1), which in most refineries is the highest percentage product yielded from crude, the reformer is the "swing" unit for making octane number. The reformer may be called on to raise the octane number of gasoline. It may also have to adjust to changing feedstock quality. The reformer may have to operate at charge rates ranging from 50% below design to 25% above design. It may have to charge poor-quality feedstocks, that is, very difficult to raise octane number, or rich feedstocks, very easy to raise octane number. If the refinery is in a hydrogen-short situation, the reformer may be called on for more hydrogen at the expense of reformate yield.

To meet these widely varying demands, the reformer catalyst responds to changes in unit operating conditions. These changes in operation which affect reformate yield and quality are termed *process variables*. The process variables having the greatest effects on reforming are:

- Feedstock properties
- Reaction temperature
- Space velocity
- Reaction pressure
- Hydrogen-to-hydrocarbon ratio (H:HC)
- Catalyst type

These variables are frequently used in discussing unit operations. An explanation of each with respect to reforming is in order.

FEEDSTOCK PROPERTIES

Feedstock properties are listed first because the effects of all other variables are related to feedstock. For example, if a reforming technologist is asked about yield or octane number or cycle life, the first question is, "What is the feedstock like?" Prior to the crude oil crisis of 1974, feedstock properties were not considered to vary significantly because each refinery could depend on a relatively constant source of crude oils.

In the 1980s, refineries have been revamped to handle a wide variety of crude types to take advantage of the economics of worldwide changes in crude supply and price. Some refiners with residuum hydrotreaters claim the ability to charge 85% of the world's crude.[1]

Not only crude oils but also naphthas, gas oils, and residual stocks are available in large quantities in the petroleum marketplace. It is possible for a reformer to be charging a feedstock rich in naphthenes one day and a few days later to be charging a feedstock low in naphthene content.

An excellent summary of the properties of fractions in the world's crude oils, including the naphthene and aromatic content of naphthas, has been published by the *Oil & Gas Journal*.[2] Naphtha properties are printed in Appendix IX.

Fig. 2–3 shows that the volume percent reformate yield depends on the naphthene and aromatic content of the reactor charge, expressed as napthenes plus two times aromatics, in liquid volume percent, or (N + 2A) liquid vol %. Reactor charge properties from commercial reforming units are shown in Table 5–1. Feedstock A (N + 2A = 47) is the most difficult to reform and produces the lowest reformate yields

TABLE 5–1
Reactor charge properties of commercial reformers

Reactor Charge	A	B	C	D
API Gravity	58.5°	54.5°	55.0°	46.9°
Distillation, ASTM D–86	°F	°F	°F	°F
IBP	161	225	202	184
10 vol %	214	244	216	292
30 vol %	237	—	230	299
50 vol %	259	266	247	305
70 vol %	285	—	267	316
90 vol %	320	320	300	335
EP	362	370	376	386
Paraffins, liquid vol %	60	56	40	34
Naphthenes, liquid vol %	33	28	43	29
Aromatics, liquid vol %	7	16	17	37
N + 2A	47	60	77	103
Sulfur, ppm wt	<0.5	<1.0	<0.5	<0.5
Nitrogen, ppm wt	<0.5	<0.5	<0.5	<0.5

of a given octane number. Feedstock D (N + 2A = 103) is easiest to reform and produces the highest yields of reformate of a given octane number. In fact, a reformer probably could not attain as high an octane reformate when charging feedstock A as when charging feedstock D because of reactor temperature limitations.

A more reasonable comparison for Fig. 2–3 is shown by feedstocks A and B. According to Fig. 2–3, when reforming to 94 RON0, reformate yield would be about 85 liquid vol % of feedstock B and about 82 liquid vol % of feedstock A. As discussed later in this chapter, the reactor temperature required to produce 94 RON0 is lower for feedstock B (in this instance, about 15°F) than for feedstock A.

Feedstock N + 2A

The N + 2A liquid vol % value of reforming feedstock is commonly used as a rule of thumb by which to judge reformer feeds. It is often referred to in technical articles and the NPRA Q&A sessions on reforming. The only known published explanation for its use as feed-quality parameter was an answer at the 1969 NPRA Q&A.[3] According to Allard of UOP, the N + 2A as calculated by the UOP equation is a function of the characterization factor, or UOP K, of the naphtha and its ASTM 50% point. The average naphthene has a K approximately one number below the average paraffin; the average aromatic has a K approximately two numbers below the average paraffin. Therefore, the aromatic contributes about twice as much as the naphthene to the reduction in K. For this reason, a correlation can be made relating the K to the N + 2A value.

Allard indicates the agreement of the calculated N + 2A from the characterization factor and 50% point as compared to the UOP laboratory-determined value shows a standard deviation of 3.0 (N + 2A). The implication is, that since the UOP characterization factor is widely used to characterize petroleum fractions, the K or, alternatively, N + 2A could be used to characterize reforming feedstocks.

The best reason for using N + 2A is that it works. The UOP equation referred to by Allard is proprietary. The equation calculates the N + 2A of a naphtha from its API gravity and ASTM distillation. The equation was tested on various naphthas and found to generally predict N + 2A within five numbers of the N + 2A PONA* analysis. One exception was an alkylate which was 100% paraffin. The equation calculated a −13

---

*PONA is standard laboratory analysis for paraffins, olefins, naphthenes, and aromatics in a liquid.

N + 2A. This confirms a truth that should be applied to not only the N + 2A but also other correlations in this chapter. These correlations should be used only for the range of conditions for which they apply. Extrapolations can lead to significant error.

Feedstock Boiling Point Range

One feedstock property which a refiner can control is boiling point range. In a modern refinery process scheme, usually the naphtha end point is controlled on the crude unit fractionator, and the initial boiling point of reformer charge is controlled on the naphtha hydrotreater fractionator. One reason for this is that a small amount of hydrocracking generally occurs in the hydrotreater, producing low-boiling-point hydrocarbons which must be removed from the naphtha before it is charged to the reformer reactors.

For a motor fuel reformer, a refiner will generally control the initial boiling point (IBP) to include in the feed those components which will reform to benzene. The boiling points of key naphthenes, aromatics, and paraffins are shown in Table 5–2. The benzene-formers are methylcyclopentane, which boils at 161.3°F, and cyclohexane, which boils at 177.3°F. Therefore, reformer reactor charge usually has an ASTM IBP of 150–160°F. Though n-pentane isomerizes to isopentane, any isopentane in the feed passes through unchanged.

Refiners do not normally find it economical to use up reformer charge capacity by including pentanes in the reactor charge. The general opinion of refiners at the 1977 NPRA Q&A session was that lowering IBP decreases octane of the reformate unless reactor temperature is raised to compensate.[4]

The final boiling point, ASTM distillation end point (EP), of the reactor charge is of interest to refiners. As explained in chapter 4, hydrocarbons boiling above 400°F are known to form polycyclic aromatics at reforming conditions. The polycyclic aromatics are responsible for carbon laydown on catalyst and short cycle life. Therefore, refiners usually specify a maximum ASTM 400°F end point on reformer reactor charge.

In the 1978 NPRA Q&A session on refining Dean Edgar estimated that, at 400°F final end point (FEP), cycle length decreases 0.9–1.3% per each 1°F increase in feed end point, and around 420°F FEP the decrease in cycle length is 2.1–2.8% per each 1°F increase in feed EP.[5] Answering the same question, L.R. Mains gave as a general rule that a 25°F increase in feed end point costs about 35% on catalyst life between levels of 375°F and 425°F feed end point. Some refiners hold reactor

TABLE 5–2
Boiling point of hydrocarbons

Hydrocarbon	Formula	Boiling Point, °F at 14.7 psia
Naphthenes		
Methylcyclopentane	C_6H_{12}	161.3
Cyclohexane	C_6H_{12}	177.3
Ethylcyclopentane	C_7H_{14}	218.2
1 cis-2-Dimethylcyclopentane	C_7H_{14}	211.2
Methylcyclohexane	C_7H_{14}	213.7
Cycloheptane	C_7H_{14}	245.8
Ethylcyclohexane	C_8H_{16}	269.2
1,1-Dimethylcyclohexane	C_8H_{16}	247.2
1,1,2-Trimethylcyclohexane	C_9H_{18}	293.4
1,2,4-Trimethylcyclohexane	C_9H_{18}	286.2
n-Butylcyclohexane	$C_{10}H_{20}$	357.7
Aromatics		
Benzene	C_6H_6	176.2
Toluene	C_7H_8	231.1
O-xylene	C_8H_{10}	291.9
M-xylene	C_8H_{10}	282.4
P-xylene	C_8H_{10}	281.0
Paraffins		
n-Pentane	C_5H_{12}	96.9
iso-Pentane	C_5H_{12}	82.1
n-Hexane	C_6H_{14}	155.7
2-Methylpentane	C_6H_{14}	140.5
n-Heptane	C_7H_{16}	209.2
2,2-Dimethylpentane	C_7H_{16}	174.6
n-Octane	C_8H_{18}	258.0
2,2-Dimethylhexane	C_8H_{18}	224.3
2,2,3-Trimethylpentane	C_8H_{18}	229.7
n-Nonane	C_9H_{20}	303.4
2,2-Dimethylheptane	C_9H_{20}	270.8
n-Decane	$C_{10}H_{22}$	345.4
n-Undecane	$C_{11}H_{24}$	384.6
n-Dodecane	$C_{12}H_{26}$	421.3

charge well below 400°F FEP to avoid accidental inclusion of high-boiling-point material. Another reason for caution is that the end point of the C_5^+ reformate product is higher than the end point of the feedstock. The reformate is lower IBP and higher EP than the feed. This means that higher-boiling-point hydrocarbons are present in the final reactor where operating severity is highest and coking is most likely to occur.

78 CATALYTIC REFORMING

An example from a commercial unit is as follows:

	°API	IBP, °F	10%, °F	30%, °F	50%, °F	70%, °F	90%, °F	EP, °F
Reactor Charge	54.6	203	233	246	258	278	312	371
Reformate	49.5	127	202	240	261	282	322	419

The feed was a Midcontinent straight-run naphtha of 68 liquid vol % N + 2A. Reformate was 92 RON0.

Contaminants (catalyst poisons) in reactor charge were discussed in chapter 4. In the 1980s a refiner usually has feed pretreatment as part of the reforming scheme. In this chapter, unless otherwise stated, correlations and charts are for reformer feedstocks containing no more contaminants than the maximums specified in chapter 4.

Cracked Naphthas

Olefin-containing stocks such as thermal-cracked, cat-cracked, coker, and pyrolysis naphthas are not included in reformer charge without first having been hydrotreated. In addition to being high in sulfur and nitrogen, these stocks contain substantial amounts of monoolefins and diolefins. The monoolefins and diolefins are undesirable in reformer feed for several reasons. First, in a reformer reactor, olefins hydrogenate, consuming hydrogen that might be needed for hydrotreating other stocks. Second, hydrogenation of olefins markedly lowers their octane number, which results in lower-octane reformate.

A comparison of olefin and paraffin octane numbers is shown in Table 5–3. Since olefins may range as high as 25%–40% of a cracked stock (Table 5–4), the effect on reformate octane can be significant.

TABLE 5–3
Octane numbers of olefins and paraffins

Compound	Formula	Research Octane Number Clear
1-Hexene	C_6H_{12}	76.4
n-Hexane	C_6H_{14}	24.8
2-Methyl-1-pentene	C_6H_{12}	94.2
2-Methylpentane	C_6H_{14}	73.4
1-Heptene	C_7H_{14}	54.5
n-Heptane	C_7H_{16}	0.0
2-Methyl-1-hexene	C_7H_{14}	90.7
2-Methylhexane	C_7H_{16}	42.4
2-Methyl-1-heptene	C_8H_{16}	70.2
2-Methylheptane	C_8H_{18}	21.7

TABLE 5-4
Cracked feedstocks before hydrotreating

	Cat-cracked Naphtha	Coker Naphtha
API Gravity	43°	46.7°
Sulfur, wt %	0.08	0.92
Total nitrogen, ppm wt	40	400
Paraffins, liquid vol %	22.8	30
Naphthenes, liquid vol %	13.4	
Olefins, liquid vol %	[a]25.9	38
Aromatics, liquid vol %	37.9	32
Octane Number		
RON0	87.8	78.3
MON0	79.0	—
Distillation, ASTM D-86		
IBP of feed	183°F	200°F
10%	207°F	242°F
30%	228°F	268°F
50%	242°F	296°F
70%	281°F	322°F
90%	330°F	362°F
EP	442°F	409°F

[a]Monoolefins 17.5%, Diolefins 8.4%

The third reason for not including olefins in reformer feed is their tendency to polymerize and form coke on the reforming catalyst—particularly true of the diolefins.

REACTION TEMPERATURE

The process variable most frequently used by refiners to control reformer operation is *reaction temperature*. By simply raising or lowering heater outlet temperatures, a refiner can raise or lower octane number of reformate. A test run on a commercial unit produced the temperature–octane curve of Fig. 5–1. The refiner has the flexibility of varying the reformate octane number clear from 85 to more than 100, provided unit equipment is not limited by the 970 *weighted average inlet temperature* (WAIT). However, note in Fig. 5–2 that in raising octane number clear from 85 to 100, the C_5^+ reformate yield dropped from 90 to 77 liquid vol % of charge.

Each feedstock has its own temperature/octane relationship. An example of two very different feedstocks and their temperature/octane relation from pilot unit tests is shown in Fig. 5–3. For the same reactor temperature, feed X produced a C_5^+ reformate about 15 RON0 higher than feed Y.

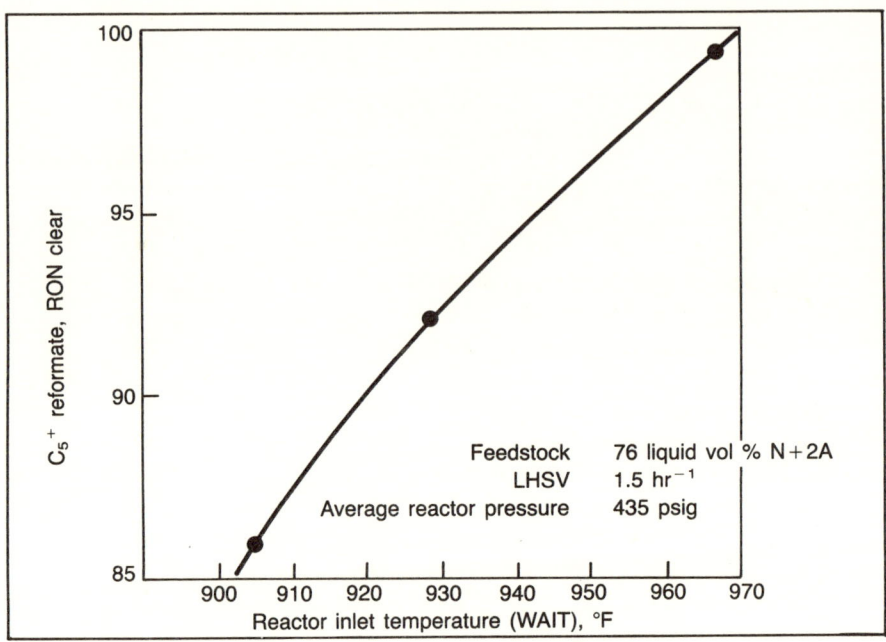

Fig. 5–1 Temperature vs octane number

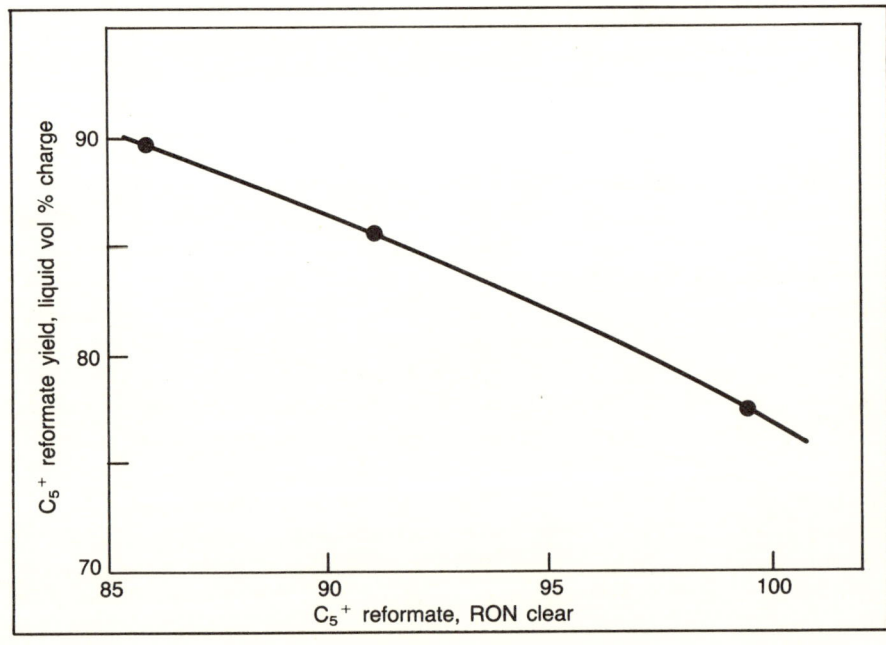

Fig. 5–2 Octane number vs yield

There is no known general correlation that can confidently show the temperature required to produce a given octane number. Licensors' estimates of reactor temperatures are probably generated from reformer kinetic models, but the results must be compared to hundreds of test runs and feedback from commercial operations to arrive at a reasonable figure.

The plant engineer is most interested in the temperature increase required to change reformate octane number. For example, in the plant test of Fig. 5–1, a temperature increase from 920°F to 943°F raised reformate octane by five numbers, or about 4.6°F-increase per octane number. As a rule, in the range of 90–95 RON0, the WAIT increase should be from 3 to 5°F per octane increase. In the 95–100 RON0 range, the WAIT increase may be from 5 to 7°F per octane.

In commercial operations, reaction temperature cannot be obtained because temperature continually changes as reactants flow through the catalyst bed. Therefore, the temperatures commonly utilized for reforming correlation are the average reactor inlet temperature and the average bed temperature.

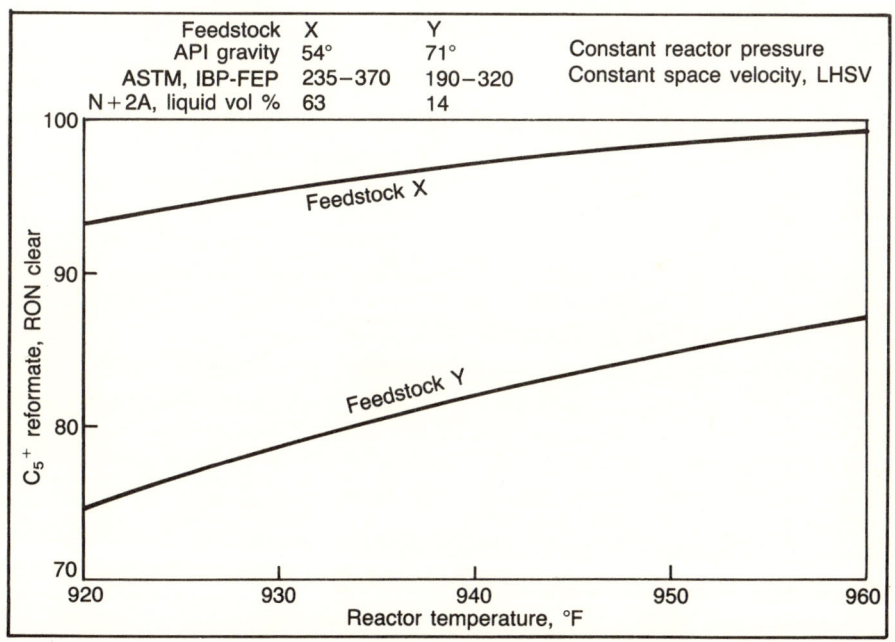

Fig. 5–3 Reactor temperature vs RON clear

82 CATALYTIC REFORMING

Since reformers are designed with three or more reactors in series and since each reactor may contain a different quantity of catalyst (chapter 3), weighted average inlet temperature is customarily used. The WAIT is the sum of the inlet temperature to each reactor multiplied by the weight percent of total catalyst in each reactor. For example, Table 5–5 shows the catalyst loading and operating temperatures from a commercial reformer. The WAIT of 904°F is the sum of the weight percent catalyst in each reactor multiplied by the inlet temperature.

TABLE 5–5
Weighted average reactor temperatures[a]

Reactor Number	1	2	3	4	Total
Inlet Temperature, °F	914	910	907	900	
Outlet Temperature, °F	772	838	876	894	
Delta T, °F	142	72	31	4	249
Average Bed Temperature, °F	843	874	892	898	
Catalyst Distribution, wt %	11	17	17	55	100
WAIT, °F					904
WABT, °F					887

[a]Commercial data

Also shown in Table 5–5 is the *weighted average bed temperature* (WABT), which is sometimes used instead of, or in conjunction with, WAIT to evaluate catalyst activity. The WABT is the sum of the *average* of the inlet and outlet temperatures of each reactor multiplied by the weight percent of the catalyst in the reactor.

In day-to-day plant operation, the WAIT is easier to observe than the WABT. Because catalyst loadings tend to be constant in any one unit, persons assigned to the unit develop a "feel" for catalyst performance from observing WAIT. WABT is frequently used in pilot unit studies where catalyst type and volume are varied as part of the studies.

The reactor inlet temperatures in Table 5–5 decrease from the first to the last reactor. This mode of operation is a descending *temperature profile*. If inlet temperatures increase from the first to the last reactor, it is an ascending profile. When the inlet temperature to each reactor is the same, it is a level profile.

There is little agreement among reforming technologists, as to which temperature profile gives the best yield structure. This topic is discussed at nearly every NPRA Q&A refining session, unless the screening committee deletes the question on the grounds it has been thoroughly covered in recent sessions. Recent discussions of temperature profile bring forth the standard replies; there is no general agreement.[6,7] Any advantages of one profile over another are minor. The temperature profile used for commercial units is plant preference.

One other important temperature in Table 5–5 is the *delta temperature* (delta T) across each reactor. As discussed in chapter 2, reforming reactions are predominately endothermic, and the temperature of vapors at the exit of a reactor is normally less than the temperature of vapors at the inlet.

The 142°F delta T of the No. 1 reactor (Table 5–5) indicates good catalyst activity for dehydrogenation of naphthenes. However, the 772°F reactor outlet temperature is probably too low for much reforming reaction. It is doubtful that the catalyst in the bottom portion of the bed is significantly contributing to the reforming reactions. The 4°F delta T across the last reactor is typical because dehydrocyclization and hydrocracking predominate at the higher bed temperature and the larger quantity of catalyst. It is not uncommon to experience a 15–20°F temperature rise in the last reactor when starting up with fresh catalyst. The delta T across a reactor depends on the reactions' occurring as well as the H:HC ratio used. This ratio is discussed later in this chapter.

SPACE VELOCITY

Space velocity is an important reforming variable because it is interchangeable with reaction temperature. Space velocity has to do with the length of time of contact between the reactants and the catalyst. Just as refiners choose an easily accessible indicator of reaction temperature in WAIT or WABT, they select a readily obtainable parameter of residence time in either *liquid hourly space velocity* (LHSV) or in *weight hourly space velocity* (WHSV).

LHSV is the volume per hour of reactor charge per volume of catalyst. The reactor charge volume is measured at 60°F and 14.7 psia, petroleum industry standards. The catalyst volume is the volume of the catalyst bed in the same units as reactor charge volume. Whether the units are barrels per hour (bbl/hr) of reactor charge per barrel of catalyst or cubic feet per hour (cu ft/hr) of reactor charge per cubic feet of catalyst, the LHSV is the same figure in hr^{-1}.

Some reformer technologists prefer to use WHSV instead of LHSV. The units are then pounds per hour (lb/hr) of reactor charge per pound of catalyst. The WHSV is still in hr^{-1} but will be a different number from LHSV.

Table 5–6 illustrates the space velocity calculation. In this instance, WHSV is a little higher than LHSV, but either will do. The LHSV is used more often because reactor charge in barrels per hour is one of

84 CATALYTIC REFORMING

TABLE 5-6
Space velocity

Reactor Charge	1,000 bbl/hr
Reactor Charge at 55°API (265.3 lb/bbl)	265,300 lb/hr
Total Catalyst	224,000 lb
Catalyst Average Bulk Density	40 lb/cu ft
Catalyst Volume	5,610 cu ft = 1,000 bbl

$$\text{LHSV} = \frac{1,000 \text{ bbl/hr}}{1,000 \text{ bbl}} = 1.0 \text{ hr}^{-1}$$

$$\text{WHSV} = \frac{26,530 \text{ lb/hr}}{224,000} = 1.18 \text{ hr}^{-1}$$

the flow rates recorded at regular intervals on reformer unit data log sheets. LHSV is a convenient and proven parameter of contact time in reforming reactors.

The reactor naphtha charge, mixed with recycle hydrogen, passes through the reactors in the vapor phase at temperatures and pressures well above the standard conditions used for LHSV. In a modern reformer, operating at 1.0 hr^{-1} LHSV, 950°F, 200 psig, and 4:1 H:HC mol ratio, total residence time through the catalyst is less than 25 seconds (sec).

Space velocity, because of its relation to length of time of contact with catalyst, affects the severity of reforming. The higher the LHSV, the greater the volume of naphtha charge per hour over a given amount of catalyst. Therefore, contact time with catalyst is less and reforming severity is lower. Conversely, lower space velocity results in higher reforming severity.

A refiner, then, has two ways to change severity of reforming: reactor inlet temperature (WAIT) and reactor charge rate (LHSV). If a refiner needs higher-octane reformate, reactor inlet temperatures can be raised as shown in Fig. 5-1. If unit limitations do not allow a temperature increase, a boost in octane number can still be achieved by lowering space velocity, which means reducing reactor charge rate.

An example is the pilot unit data of Table 5-7. A 232°F IBP to 393°F FEP Midcontinent naphtha was reformed over a bimetallic catalyst at 200 psig and a constant 965°F reactor inlet temperature. Reformate was varied from 82.1 to 100.5 RON0 by changing LHSV from 3.0 to 1.25 hr^{-1}.

Data derived from another pilot test, using a different feedstock and a different catalyst than those of Table 5-7, is plotted in Fig. 5-4 to show the relationship of space velocity and reactor inlet temperature at constant octane number.

TABLE 5-7
Effect of space velocity on octane number (pilot unit)[a]

Space Velocity (LHSV), hr^{-1}	1.2	2.0	3.0
Reactor Inlet, °F	965	965	965
C_5^+ Reformate Octane Number Clear	100.5	90.8	82.1

[a]Feedstock: Midcontinent naphtha, 54.8°API, 232°F IBP, 393°F EP, 60 (N + 2A)

A more useful plot is Fig. 5–5, which shows the adjustment that must be made in reactor inlet temperature as space velocity is varied. Thus, with feedstock A, doubling LHSV from 1.0 to 2.0 hr^{-1} requires raising WAIT about 30°F to maintain reformate at 95 RON0. Data from commercial units indicate that, generally, for reformate octane numbers in the range of 90–100 RON0, doubling LHSV requires a 30°F increase in reactor inlet temperature.

Highly paraffinic naphthas require a 40–50°F increase in temperature as space velocity doubles. Naphthas, low in paraffins, only require a 15–20°F temperature increase as space velocity doubles. Most reforming operates in the range of 1.0–2.0 hr^{-1} LHSV.

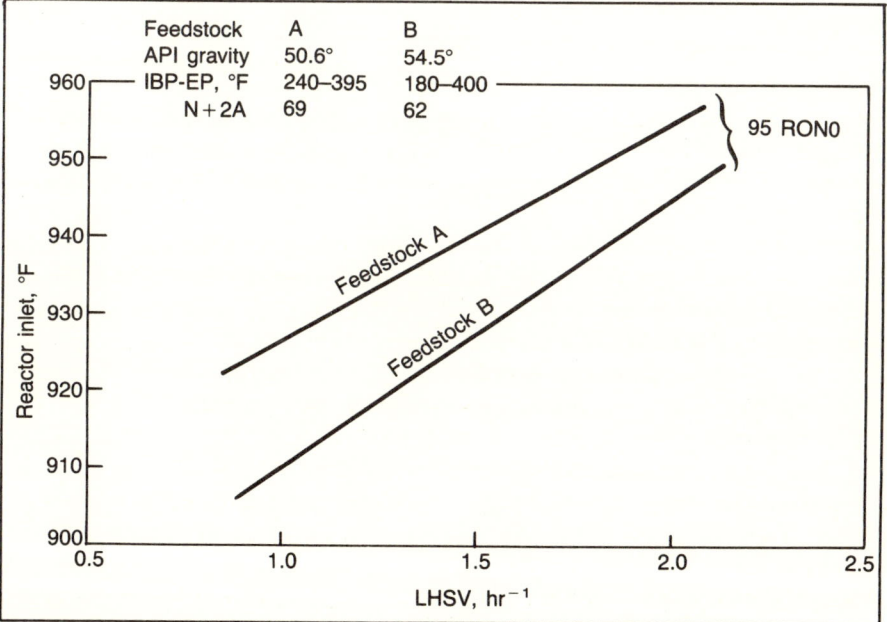

Fig. 5–4 Relationship of space velocity to reactor inlet temperature

86 CATALYTIC REFORMING

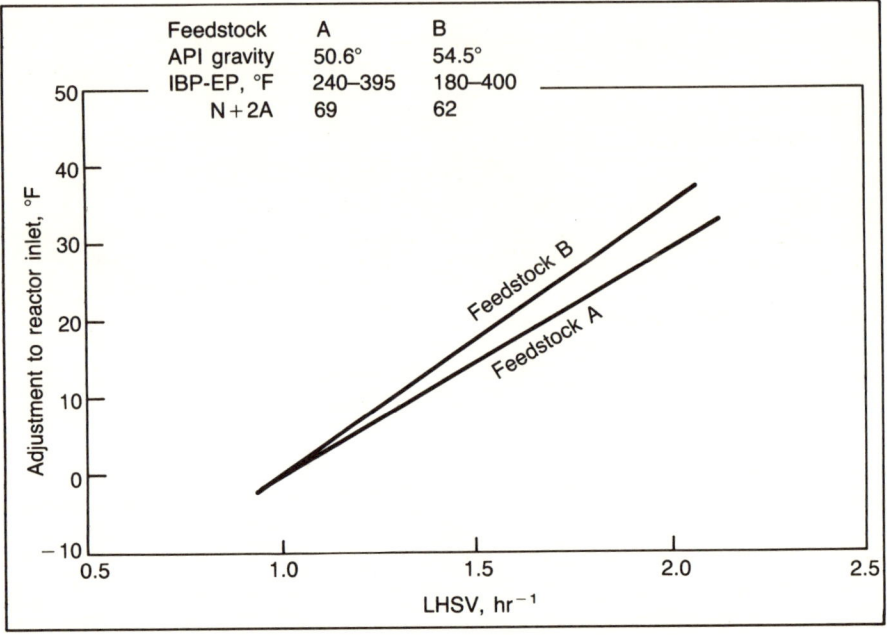

Fig. 5–5 Adjustment to reactor inlet temperature for space velocity

The space velocity/temperature relationship is sometimes useful in estimating the *maximum charge rate* to a reformer at different reformate octane numbers. As an illustration, a reformer has a reasonable cycle life at a charge rate of 20,000 b/sd (1.5 hr^{-1} LHSV) with a start-of-run WAIT of 940°F, producing 94 RON0 $C_5{}^+$ reformate. Each increase of one octane number requires a WAIT increase of 5°F. Maximum reactor inlet temperature (WAIT) at the end of the cycle is 980°F.

The following estimates are made:

Charge Rate, b/sd	LHSV, hr^{-1}	WAIT, °F	RON0
16,000	1.20	940	96.0
18,000	1.35	940	95.0
20,000	1.50	940	94.0
24,000	1.80	940	92.4
26,000	1.95	940	91.6

In this example, WAIT is held at 940°F at the start of the run because temperature is increased as the run progresses to compensate for loss of catalyst activity. The above estimates, then, are for approximately the same cycle life as the 94 RON0 base, although, as discussed later, the higher severity (96 RON0) has a shorter cycle life and the lower severity

(91.6 RON0) a longer cycle life than the 94.0 RON0 base case. At the higher octane numbers, the refiner has to decide if shorter cycles, that is, more frequent shutdowns for catalyst regeneration, are acceptable. Unit design limitations such as pump capacities and metallurgy must also be considered.

Modern commercial reformers usually operate between 1.0 and 2.0 hr^{-1} LHSV. At space velocities below 1.0 hr^{-1}, undesired side reactions and increased hydrocracking occur, reducing reformate yield. Minimum LHSV has been discussed at NPRA Q&A sessions.[8,9]

REACTION PRESSURE

The pressure under which reforming reactions take place significantly affects yields and cycle length. Decreasing pressure increases both dehydrogenation of naphthenes and dehydrocyclization of paraffins, favoring an increase in production of aromatics (higher-octane reformate) and hydrogen. The adverse effects of reduced pressure are increased catalyst coking and shorter cycle life. Higher pressures cause higher rates of hydrocracking. More hydrocracking causes a loss of reformate yield for a given octane number. Higher pressure decreases coking of the catalyst, resulting in longer cycle life.

Reformers usually operate with three or four reactors in a series, and the pressure is different in each reactor. What is really meant by operating at 200 psig or 500 psig? Reforming technologists prefer to use the average reactor pressure, if it is available, for estimating yields and cycle life. Otherwise, the average of the first-reactor inlet pressure and the last-reactor outlet pressure is employed. If that is unavailable, the average pressure is estimated from the outlet pressure of the last reactor, together with an estimated differential of about 20–30 psi from the inlet of the first reactor to the outlet of the last reactor, for most units built before 1975.

Some of the newer units have a pressure drop of only 10–15 psi. (Examples from commercial units are shown in Table 5–8.) Although these units operate at 435 psig and 260 psig, respectively, reformers are being designed in the 1980s for 50–100 psig reactor pressures.

The real incentive for reducing pressure in reformer reactors is more reformate yield, with the added benefit of increased hydrogen. The effect of pressure on yield is shown in Fig. 5–6. A refiner would be quite pleased to increase his C_5^+ reformate yield at 100 RON0 from 79.7 to 83.5 liquid vol % of charge, which, according to Fig. 5–6, would be expected from reducing the pressure from 300 psig to 100 psig.

Another interpretation of Fig. 5–6 is that, at the 100 RON0 level, yield increases or decreases about 2 liquid vol % of charge per 100 psig change in pressure. At the 90 RON0 level, yield is about 1 liquid vol %

88 CATALYTIC REFORMING

TABLE 5-8
Pressure differential across reformer reactor system

	Unit A[a]				Unit B[b]		
Reactor number	1	2	3	4	1	2	3
Inlet pressure, psig	451	447	438	427	270	260	250
Pressure differential, psig	5	1	1	6	3.8	5.4	3.9

[a]LHSV at 1.9 hr^{-1}
H:HC at 4
[b]LHSV at 1.6 hr^{-1}
H:HC at 3.2

of charge per 100 psig change in pressure. The increase in hydrogen yield is not as clearly defined, but it generally ranges from 50 to 100 scf/bbl increase per 100 psig decrease in pressure.

A reduction in pressure, while of significant benefit to reformate yield, unfortunately increases the rate of coking of the catalyst, which in turn deactivates the catalyst and shortens cycle life. The effect of increasing reforming severity either by raising octane number at constant

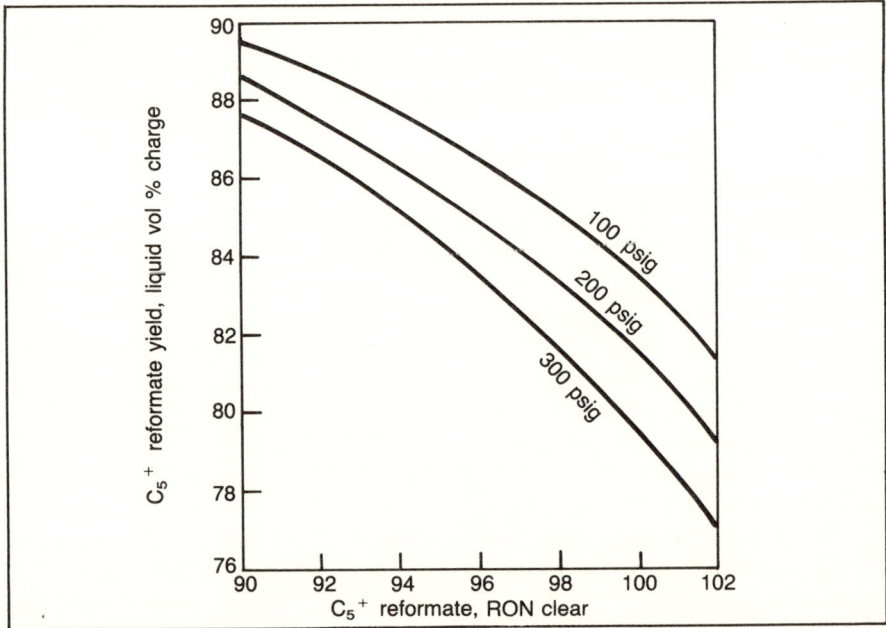

Fig. 5-6 Effect of pressure on reformate yield *(courtesy UOP Process Div.)*

pressure or reducing pressure at constant octane number or by changing both pressure and octane number is shown in Fig. 5–7.

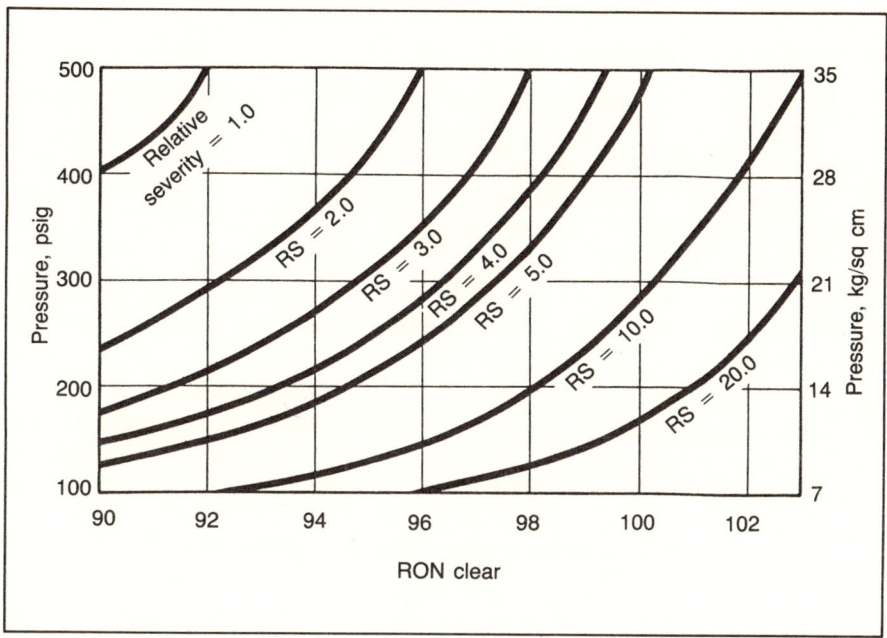

Fig. 5–7 Effect of operating pressure and octane on catalyst deactivation *(after D'Auria, Tieman, and Antos, ref. 10)*

Here the measure of catalyst deactivation rate is termed *relative severity*. For example, at 300 psig, raising research octane number clear from 92.2 to 96.5 doubles relative severity (two vs four). Thus, if the cycle life were 12 months at 92.2 RON0, it would be only six months at 96.5 RON0.

Likewise, at a constant research octane number clear of 92.2, reducing pressure from 300 to 160 psig doubles relative severity and halves cycle life. The bimetallic catalysts permit refiners to operate reformers at lower pressures and higher octane numbers.

Reactor pressure markedly affects reformate yield and coking of catalyst, but how does pressure relate to reactor inlet temperature? If a reformer is producing 100 RON0 reformate at 300 psig and 970°F WAIT, what is the WAIT required to maintain 100 RON0 if pressure is reduced to 100 psig to gain yield? To my knowledge, this aspect of process variables is missing from published literature. One set of pilot tests showed that reducing pressure from 500 to 200 psig only permitted a decrease of about 5°F WAIT to maintain a constant reformate octane number clear of 100. Commercial experience shows the same trend. Reactor pressure has a relatively minor effect on catalyst activity.

HYDROGEN-TO-HYDROCARBON RATIO

All reformers produce hydrogen. All reformers circulate hydrogen over the catalyst to protect the catalyst from rapid coking. The circulating hydrogen is the major component of a compressed gas stream *(recycle gas)* coming off the product separator downstream of the last reactor (Fig. 3–1).

The ratio of the mols of hydrogen in the recycle gas to the mols of hydrocarbon (reactor charge) is the *hydrogen-to-hydrocarbon molar ratio* (H:HC). This ratio is one of the process variables a refiner utilizes in controlling reformer operations.

An example of the calculation of H:HC ratio is:

Reactor Charge: 800 bbl/hr, 55°API, 265.3 lb/bbl, 120 mol wt

Recycle Gas: 3,350,000 scf/hr, 80 mol % hydrogen

$$\text{Hydrogen} = \frac{3{,}350{,}000 \text{ scf/hr} \times 0.80}{379 \text{ scf/mol}} = 7{,}071.2 \text{ mol/hr}$$

$$\text{Hydrocarbon} = \frac{800 \text{ bbl/hr} \times 265.3 \text{ lb/bbl}}{120} = 1{,}766.7 \text{ mol/hr}$$

$$\text{H:HC mol ratio} = \frac{7{,}071.2}{1{,}766.7} = 4.0$$

Even though the recycle gas contains 20 mol % hydrocarbons, which may include pentanes, note that these hydrocarbons are not included when determining H:HC ratio.

The main purpose of hydrogen recycle is to increase hydrogen partial pressure in the reactors. The hydrogen reacts with coke precursors, removing them from the catalyst before they can form polycyclic aromatics which ultimately deactivate the catalyst.

Reforming technologists use the H:HC ratio as a convenient expression of the relative amounts of hydrogen in the reactor. Thus, at an H:HC ratio of 8, there should be twice as much hydrogen in the reactor as at an H:HC ratio of 4.

Little information is available for comparing the coke laydown at different H:HC ratios. A recent article reported that, in bench scale tests, reducing H:HC ratio from 8 to 4 increased carbon on catalyst 1.75 times and reducing H:HC ratio from 4 to 2 increased carbon on catalyst

PROCESS VARIABLES AND UNIT OPERATION   91

3.6 times.[11] Dean Edgar suggests a reduction in H:HC ratio from 5 to 4 shortens cycle length by about 20%.[12]

Reformers designed and built in the 1950s generally operated in the range of 8–10 H:HC ratio and 400–500 psig. Reformers in the 1980s are designed to operate in the range of a 2–5 H:HC ratio at pressures of 50–150 psig.

There are two main incentives for reducing H:HC ratio. The major reason is a reduction in energy costs for compressing and circulating hydrogen. Maintaining a high H:HC ratio requires a lot of energy to operate either a steam turbine or an electrically driven compressor.

The second H:HC reduction incentive is that a lower hydrogen partial pressure favors naphthene dehydrogenation and dehydrocyclization. At least one licensor of reforming (Engelhard, chapter 7), splits hydrogen recycle so that the H:HC ratio in the first reactor is lower than in the following reactors. Through experience, each refiner determines which H:HC ratio best fits his reforming needs.[13,14]

CATALYST TYPE

Refining catalysts are available in a variety of contained metals,* metals concentrations, shapes, sizes, and densities (chapter 4). Before 1970 a refiner was quite limited in his selection of catalyst. In the 1980s licensors are more willing to sell their catalysts for use in nonlicensed units. It is not unusual to use a catalyst in the first reactor that differs from the one in the last reactor. This variation sometimes improves yields and sometimes reduces investment in precious metals.

The high-severity operation of reformers in the 1980s requires short cycle times, six months or less for some units. The frequent shutdowns provide opportunities to change the catalyst. Although a change of catalyst type is an option more easily chosen now than in the past, the decision for a change should be made from pilot tests and estimates from catalyst suppliers. In the future, catalyst type will probably be used more and more as a process variable.

MONITORING REFORMER OPERATION

Catalytic reforming units operate best when the process variables are adjusted to produce the maximum yield of desired products within the limits of acceptable catalyst cycle life.

Maintaining a reformer at its best operating level is the responsibility of the plant engineer assigned to the cat-reformer unit. His obligation is to follow day-to-day operation, to watch for signs that indicate a need

---

*If a catalyst contains platinum and iridium, for example, they are the contained metals.

92 CATALYTIC REFORMING

for changes, and to advise the operations division of any necessary changes. The engineer should be present during regenerations and turnarounds. He should keep abreast of new developments in catalytic reforming and make long-range studies of changes that could make the reformer a better unit.

Material Balance

The primary concerns of the reforming technologist are reformate yield, reformate quality, and catalyst performance. The starting point is a unit *material balance,* a weight balance that usually includes data not on the daily schedule. If the metered flow rates and laboratory inspections of charge and yield do not give a good weight balance, there is little confidence in conclusions drawn from the data. The best weight balance is made when the sums of the streams to and from the unit equal 100 wt %.

Such accuracy is seldom achieved in commercial practice or, for that matter, in pilot unit tests. Even the best orifice flow meters have an accuracy of ±2%, and turbine meters have an accuracy of ±0.5%.

For judging commercial reformer performance, if the material balance is in the range of 98–102 wt %, the yield data can be used after adjusting the charge or reformate volumes to bring the balance to 100 wt %. Any balance less than 98 wt % or greater than 102 wt % is suspect, and the data must be viewed with suspicion.

The frequency of running a material balance for the reformer varies with each refiner.[15] One material balance a week should be the minimum number, and two or more a week are better. The benefit of frequent material balances, other than better yield data, is that other data related to unit performance are also obtained.

An example of a material balance from a commercial unit is shown in Table 5–9. Some data have been deleted to maintain confidentiality. All of the process variables discussed at the beginning of this chapter are included in Table 5–9, with the exception of the kind of catalyst.

The fractionator on the naphtha hydrotreater is the type utilizing a centerwell tray. The reformer reactor charge is drawn from the centerwell. The material balance based on metered flow rates is 100.42 wt %. In this program, the reactor charge is adjusted to close the weight balance. On this particular day, the reactor charge was arbitrarily raised by 67 b/sd to make a 100% weight balance. Some prefer to close the balance by adjusting the reformate yield instead of reactor charge. The rest of the streams are a small percentage of the total yield and would have to be changed a proportionately large amount to affect the balance.

TABLE 5-9
Catalytic reformer material balance (Charge adjusted to 100% wt balance)

Prefractionator Section Balance

	b/sd	% Prefractionated Feed	
		Vol %	Wt %
Prefractionator Feed (Naphtha HDS Stabilizer Bottoms)	22,005.52		
Prefractionator Overhead Liquid	2,647.64	12.03	10.53
Prefractionator Bottoms	155.09	0.70	0.80
Reactor Charge	15,965.91	72.55	73.89
Total Side Draw	15,965.91	72.55	73.89
Total Prefractionator Yield	18,768.64	85.29	85.22

Separator Gas

Yields	Mol %	Mscfd	b/sd
H_2	82.38	10,569	
$N_2 + O_2$	0.06	8	
C_1	8.14	1,044	
$C_2 + C_2^*$	4.39	563	
H_2S	0.00	0	
C_3	2.85	366	240
C_3^*	0.00	0	0
iC_1	1.04	133	104
nC_1	0.61	78	59
iC_5	0.23	30	26
nC_5	0.11	14	12
C_6^+	0.19	24	22
Totals	100.00	12,830	462

Reformer Section Material Balance

	Stabilizer Gas			Stabilizer Liquid		Reformate		Total Yield		% Charge	
Yields	Mol %	Mscfd	b/sd	Liquid Vol %	b/sd	Liquid Vol %	b/sd	Mscfd	b/sd	Liquid Vol %	Wt %
H_2	21.82	191						10,760			1.37
$N_2 + O_2$	0.06	1						8			0.02
C_1	12.17	106		0.00	0			1,151			1.17

TABLE 5-9
Catalytic reformer material balance (Charge adjusted to 100% wt balance)—cont'd

Reformer Section Material Balance—cont'd

Yields	Stabilizer Gas			Stabilizer Liquid		Reformate		Total Yield		% Charge	
	Mol %	Mscfd	b/sd	Liquid Vol %	b/sd	Liquid Vol %	b/sd	Mscfd	b/sd	Liquid Vol %	Wt %
$C_2 + C_2^*$	25.52	223		7.77	106	0.00	0	963			1.83
H_2S	0.00	0		0.00	0	0.00	0	0			0.00
C_3	25.27	221	145	32.07	437	0.05	7		822	5.15	3.50
C_3^*	0.06	1	0	0.06	1	0.00	0		1	0.01	0.00
$i-C_4$	10.00	87	68	36.54	498	1.17	155		677	4.24	3.20
$n-C_4$	4.94	43	32	23.20	316	0.00	0		562	3.52	2.76
C_4^*	0.15	1	1	0.38	5	5.39	714		6	0.04	0.03
$i-C_5$	0.02	0	0	0.10	1	3.50	464		741	4.64	3.89
$n-C_5$	0.01	0	0	0.00	0				476	2.98	2.52
C_6^+	0.02	0	0	0.11	2	89.89	11,907		11,927	74.71	79.69
Totals	100.00	875	247	100.00	1,363	100.00	13,246	12,882	15,212	95.28	100.00

Total Reactor Products Equal 100.42 Wt % of 15,899 b/sd Metered Reactor Charge
100.00 Wt % of 15,966 b/sd Adjusted Reactor Charge

Laboratory Analysis

	°API (Specific Gravity)	ASTM Distillation (Altitude Correction Applied)							Color	Rvp	Flash	Mol Wt	High Heating Value BTU/cu ft
		IBP	10%	30%	50%	70%	90%	EP					
Naphtha HDS Charge	64.6	106	147	195	236	276	330	374	30.0	0.0			
Prefractionator Charge	62.0	110	161	201	235	271	317	330	30.0				
Prefractionator Bottoms	39.4								0.3		0		

PROCESS VARIABLES AND UNIT OPERATION

Prefractionator Overhead Liquid	89.5										18.5	
Reactor Charge	58.5	144	204	225	251	281	320	397			113.62	
Reformate	50.6	110	161	203	239	276	323	407	30.0	16.0 (Chromatograph)		5.0 / 5.0
Separator Gas	(0.2393)											6.93
Stabilizer Gas	(1.0371)											30.03
Stabilizer Overhead Liquid	132.4433									172.40		50.65
												510.7 / 1651.9

PONA Analysis Vol %

	Paraffin	Olefin	Naphthene	Aromatic	Calculated N + 2A	RON Clear
Reactor Charge	0.0	0.0	0.0	0.0	50.72	
Reformate	0.0	0.0	0.0	0.0		91.0

Process Data

Reactor Data	No. 1	No. 2	No. 3	No. 4	Total
Reactor Inlet, °F	925	926	922	923	
Delta Temperature, °F	105	38	18	4	165
% of Total Delta T	63.64	23.03	10.91	2.42	100.00
Average Temperature, °F	872.5	907.0	913.0	921.0	
Weighted Average Inlet Temperature, °F					923.8
Inlet Pressure, psig	463	419	378	400	
Delta Pressure, psi	2	3	4	8	
Corrected Total Delta T 8:1 H:HC + 90% H_2					149
Space Velocity, bbl charge/hr/bbl catalyst					1.21
Catalyst Life, bbl/lb					100.91
Total Catalyst Volume, bbl			Total Catalyst, lb		

TABLE 5-9
Catalytic reformer material balance (Charge adjusted to 100% wt balance)—cont'd

Operating Conditions

	Temperature, °F	Pressure, psig	Flow Rate, Mscfd	Net Heating Value BTU/cu ft	scf/bbl Reactor Charge	Mol % H_2	H:HC Molar Ratio
H_2 to Naphtha HDS Unit	147	470	8,812				
H_2 to Distillate HDS Unit	147	470	4,017				
H_2 to Fuel	86	349	0				
Total Separator Gas			12,830	511	804		
Stabilizer Gas	71	167	875	1652	55		
Recycle Gas	147	470	79,218	511		82.38	4.7

Other Yield Data

Total Propane Yield Liquid Vol % Reactor Charge	5.15	
Total Butane Yield Liquid Vol % Reactor Charge	7.76	
Total Propane + Butane Yield Liquid Vol % Reactor Charge	12.91	Chloride Water Injection
Total Pentane + Yield Liquid Vol % Reactor Charge	82.33	
Stabilizer Accumulated Liquid Vol % Reactor Charge	8.54	
Stabilizer Accumulated Liquid scf/bbl Charge	119.956	Water Injection Rate, ppm 5.0
Stabilizer Accumulated Liquid lb/bbl	187.53	Chloride Injection Rate, ppm 2.0
Stabilizer Overhead Gas and Liquid Scf/bbl Charge	175	
C_4-Stabilizer Overhead Gas and Liquid scf/bbl Charge		

Utilities

		Yield Octane Data	
Cooling Water Temperature, °F		C_5^+ RON Clear	90.9
4 A.M.	49		
4 P.M.	54		
Fuel Used MBTU/bbl Reactor Charge	280.87	Reid Vapor Pressure	4.40
Net Steam Used M lb/bbl Reactor Charge	0.031	C_5 Plus Yield Liquid Vol % Reactor Charge	82.33
		Corrected C_5^+ Fresh Catalyst Yield	81.78
		Yield Deviation	0.55

A PONA analysis was not run on the reactor charge on this day, and the N + 2A of 50.72 liquid vol % is a calculated value. The separator gas and recycle gas have the same composition, so only the chromatographic analysis of the separator gas is shown. The volume of the recycle gas is listed under the section subtitled Operating Conditions. In the Process Data section, one item needs explanation—the corrected total delta T.

In the adiabatic reactors, the endothermic reforming reactions cause a drop in temperature of the total flowing stream. The recycle gas, which is combined with reactor charge and passes through the reactors, acts as a heat sink. The higher the hydrogen-to-hydrocarbon ratio, the less the delta T for constant severity. The specific heat (cp) of hydrogen at constant pressure is 3.41 BTU/lb/°F at 60°F is high compared to the specific heat of hydrocarbons (0.26 for toluene and 0.385 for n-heptane). For this reason, correlations were developed so that the observed delta T could be corrected to a standard H:HC ratio of 8 and a hydrogen concentration of 90 mol %.

As calculated in the printout, the observed delta T of 165°F at an H:HC ratio of 4.7 and at 82.38 mol % hydrogen would have been only 149 at an H:HC ratio of 8 and at 90 mol % hydrogen. Adjustments of this kind to a standard condition help in judging day-to-day operations.

One of the most important items in the printout is the *yield deviation,* which on this day was +0.55 liquid vol % of reactor charge. This deviation compares the C_5^+ reformate yield on the day of the printout with the C_5^+ reformate yield expected if fresh catalyst were in the reactors. The fresh catalyst yield is obtained from a correlation similar to Fig. 2–3 and is then corrected for pressure and N + 2A.

The difference between the corrected fresh catalyst yield and the observed yield is the yield deviation. On the day of Table 5–9, the yield deviation of +0.55 liquid vol % indicated the catalyst had good selectivity and performed as well as would be expected from fresh catalyst.

Other yields of special interest are total separator gas, as standard cubic feet per barrel (scf/bbl) of charge, together with the mol % hydrogen in the gas, which, to an engineer familiar with operation of the unit, would indicate the degree of hydrocracking. Another indication of hydrocracking is the stabilizer overhead gas and liquid; these streams are mostly butanes and lighter hydrocarbons, produced by hydrocracking.

Butane and Lighter Yields

Although C_5^+ reformate yield and octane number are of primary importance, the yields of n-butane, isobutane, propane, ethane, meth-

98 CATALYTIC REFORMING

ane, and hydrogen are also significant to a refiner. Furthermore, these yields must be determined to obtain an overall material balance.

With increasing reforming severity, C_5^+ reformate yield decreases and the yield of butane and lighter hydrocarbons increases. An example of the magnitude of these yields with change in reforming severity (change in the octane number) is shown in Fig. 5–8, the result of pilot unit tests. The yield of each of the butane-and-lighter hydrocarbons increases with increasing reformate octane number.

These light hydrocarbons are generally of lower market value than gasoline, so a refiner prefers to maximize the yield of C_5^+ reformate and minimize the yield of C_4-and-lighter hydrocarbons. In certain circumstances, such as short hydrogen supply or high-priced liquefied petroleum gas (LPG), a refiner may reform to a higher-octane-number reformate than needed to obtain more hydrogen, propane, or butane.

Because of the hydrogenation activity of reforming catalyst, adding olefins (propylene or butylenes) to reformer feed to produce propane or butane for LPG has been suggested. A U.S. patent covering such a process showed that, if 25 vol % propylene is added to reformer feed, the bulk of the propylene appears as propane in the product.[16] Adding n-butane to reformer feed to produce isobutane by isomerization in the reformer reactors has also been proposed. Increased isobutane yield would be realized but would probably not be justified because of a reduction in unit naphtha charge capacity.

Sometimes a reformer technologist must estimate reforming yields on various feedstocks. The C_5^+ reformate yield is generally obtainable from published N + 2A correlations, but correlations for the C_4-and-lighter-hydrocarbons yields are missing.

For making "ball park" estimates, the curves of Fig. 5–9, are useful. The chart gives a 100% weight balance because hydrogen is obtained by difference. The correlation of Fig. 5–10 is not extremely accurate, but it is useful for calculating differences between levels of reforming severity.

Plotting Daily Operating Data

A material balance plus key operating data (Table 5–9) gives a reasonable picture of reformer unit performance on the day the data were taken. However, reforming catalyst is sensitive to subtle changes in operating conditions and variations in feedstock properties. For this reason, a continuing plot of key data and process variables is used by nearly all reformer technologists to monitor unit performance.

Examples of reforming data plots are shown in Figs. 5–10 through 5–14. The data are not from the same unit as that of Table 5–9. The

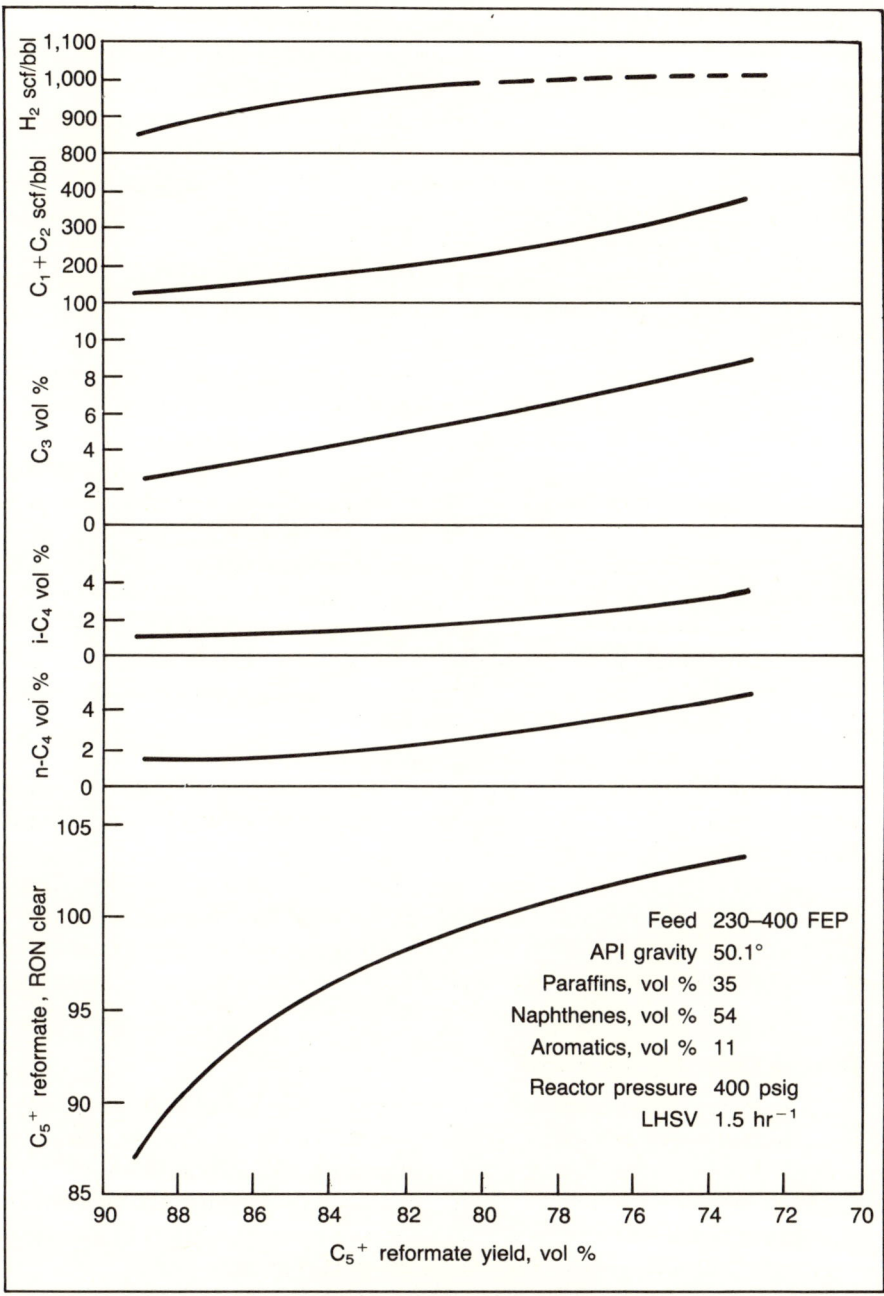

Fig. 5–8 Reforming yields

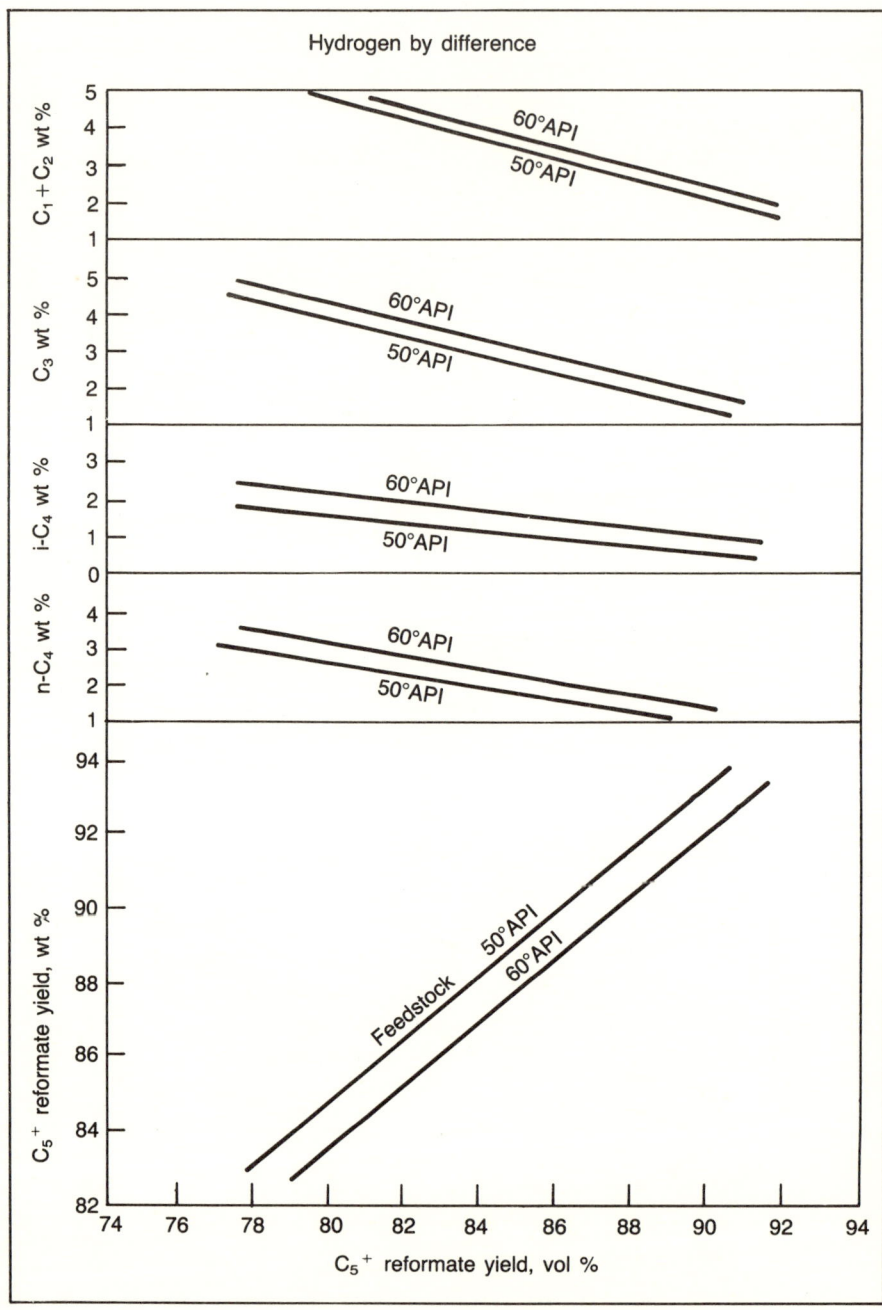

Fig. 5–9 Estimating C_4 and lighter yields from reforming

PROCESS VARIABLES AND UNIT OPERATION 101

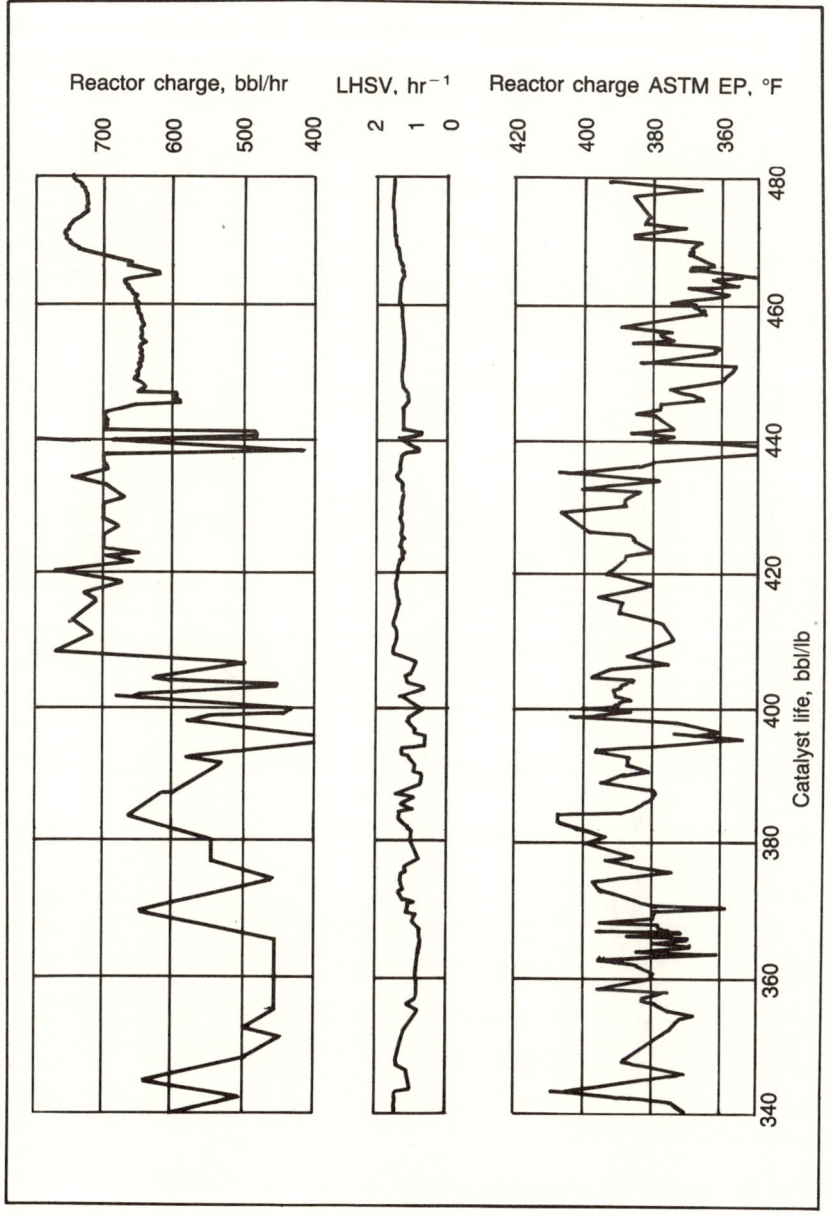

Fig. 5–10 Catalytic reformer operating plots

102 CATALYTIC REFORMING

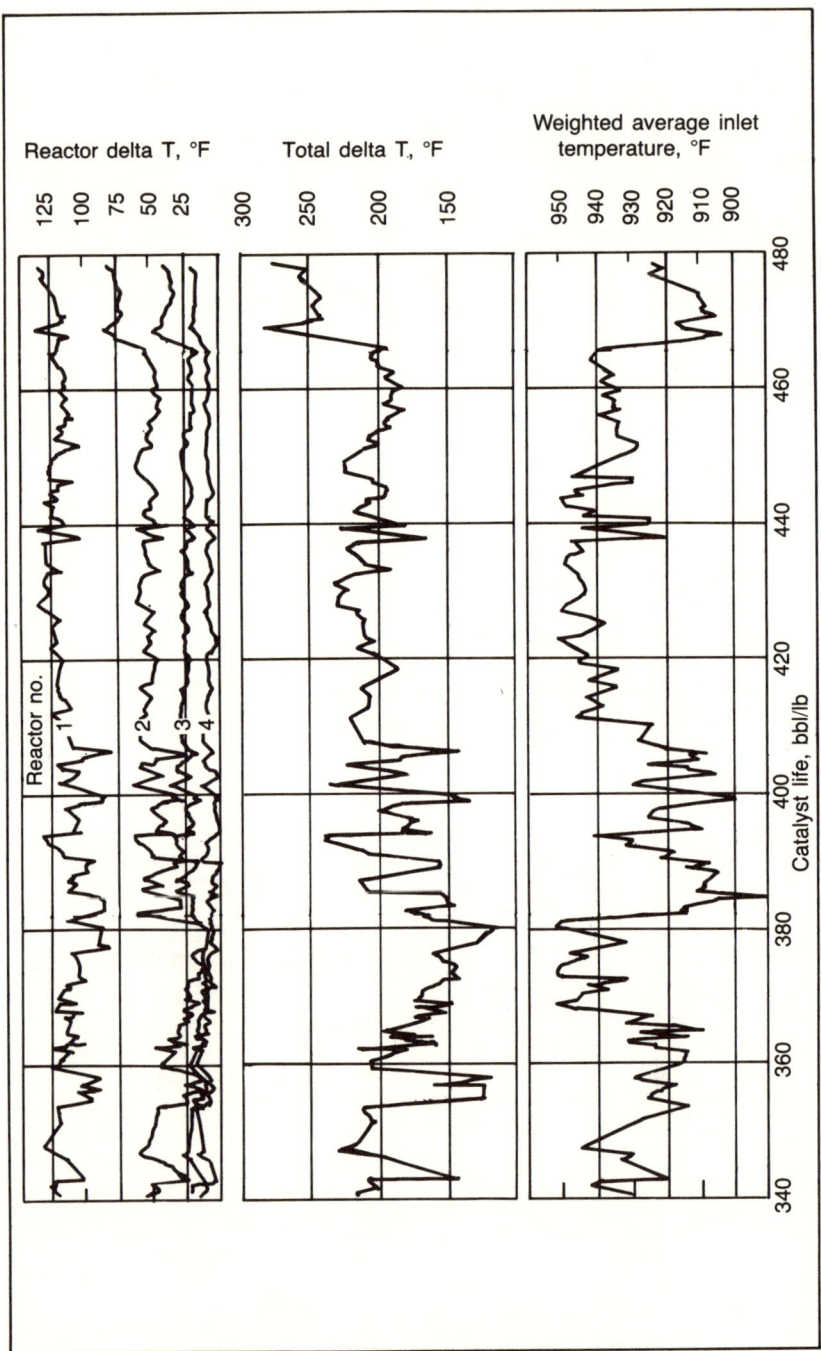

Fig. 5–11 Catalytic reformer operating plots

PROCESS VARIABLES AND UNIT OPERATION 103

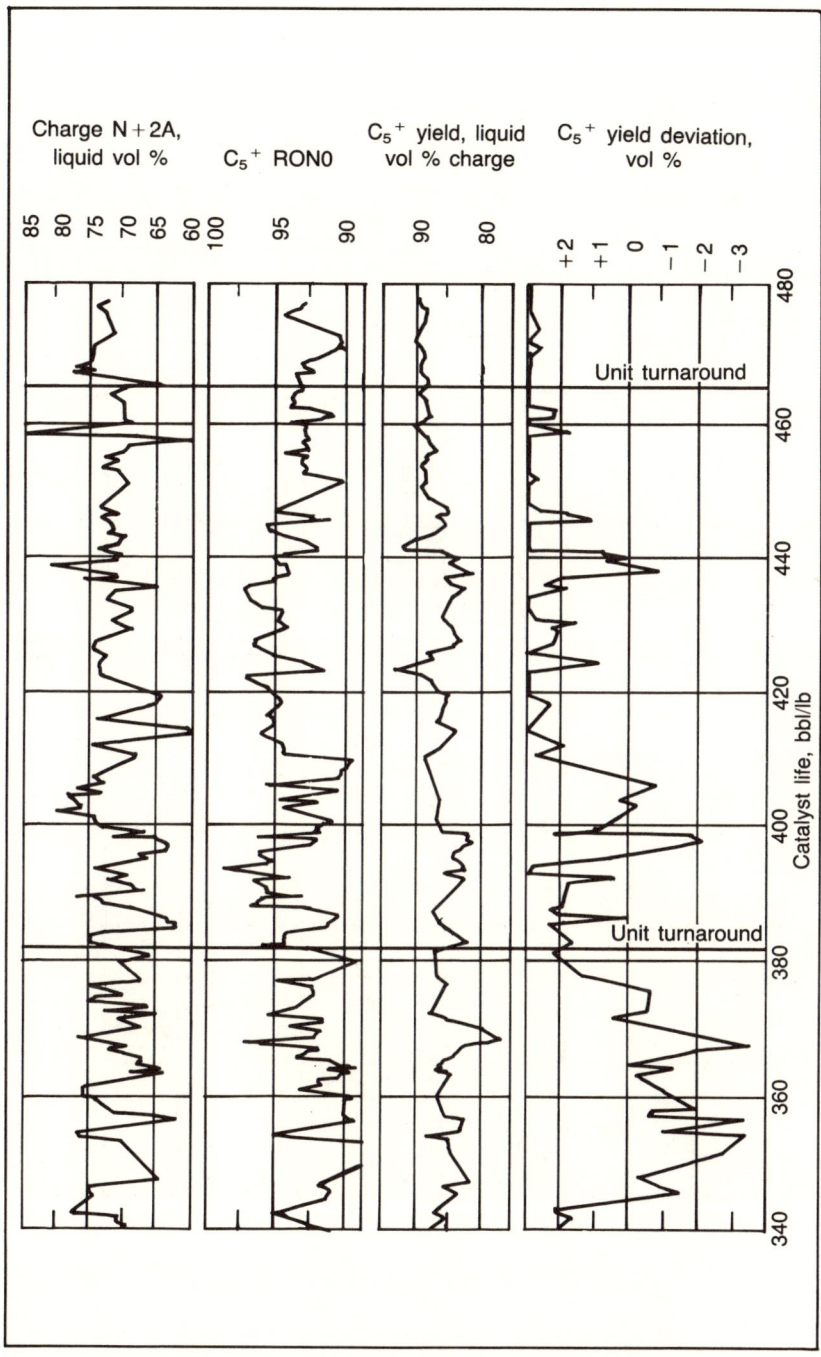

Fig. 5–12 Catalytic reformer operating plots

104 CATALYTIC REFORMING

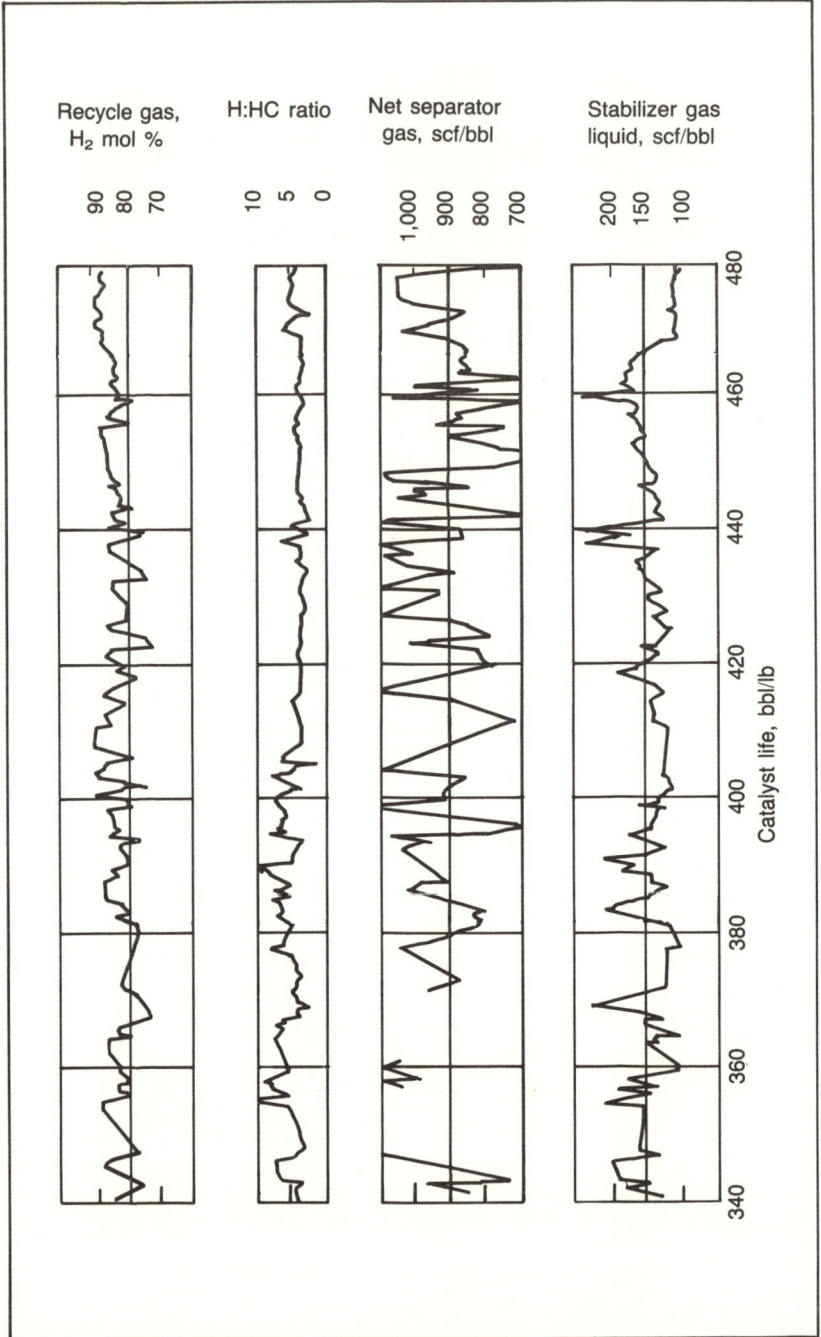

Fig. 5–13 Catalytic reformer operating plots

PROCESS VARIABLES AND UNIT OPERATION 105

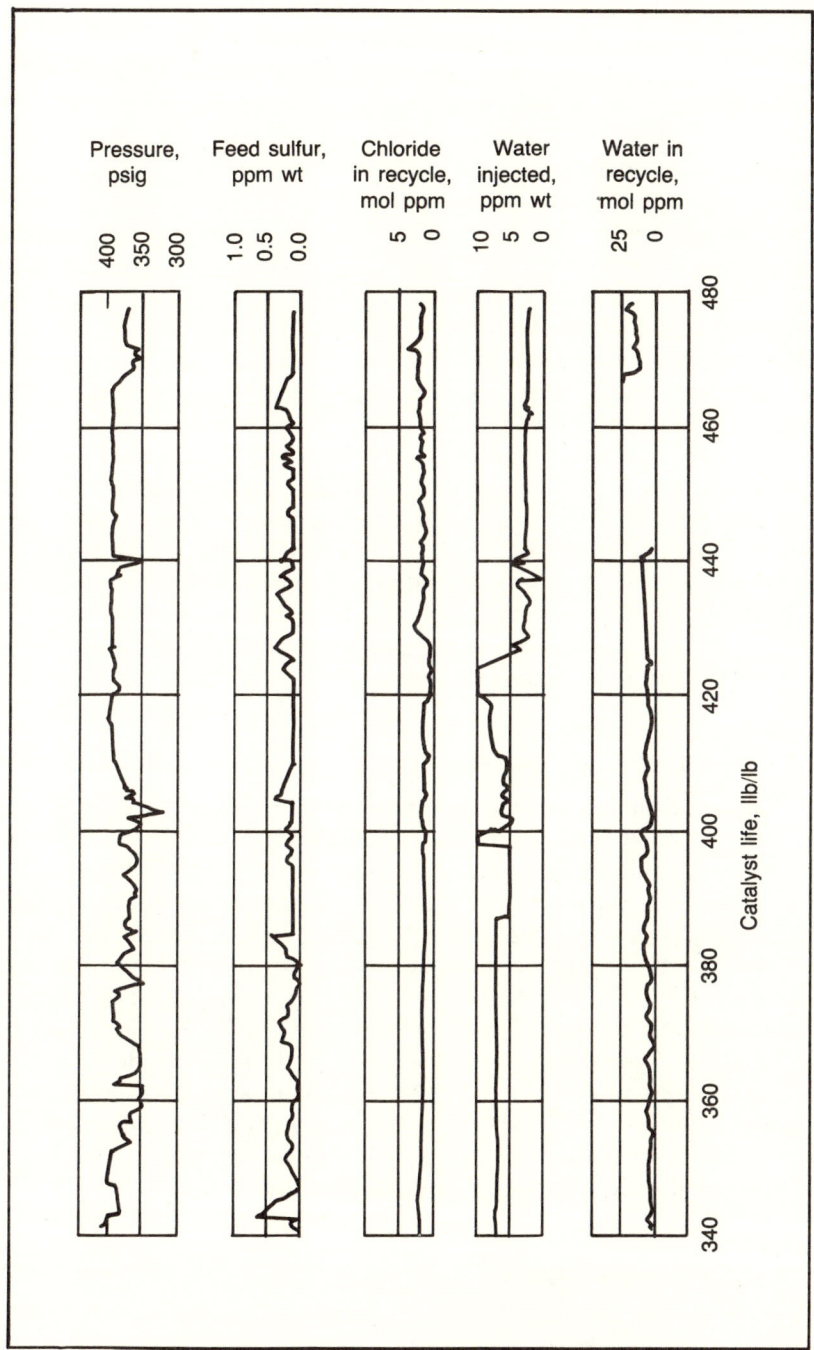

Fig. 5–14 Catalytic reformer operating plots

plots cover a period of six months, or 140 bbl/lb, of catalyst life. As noted on Fig. 5–12, the unit was shut down for catalyst regeneration at 380 bbl/lb and at 465 bbl/lb. Thus, the catalyst *cycle life* was 85 bbl/lb.

The plots illustrate the variations normally experienced in the operation of a commercial catalytic reformer. The swings are typical of operating units, not only catalytic reformers, but also other refinery units such as cat crackers and hydrotreaters.

The plant engineer looks for trends. If one point shows an unusually large rise or fall, he does not jump to conclusions, but waits until the next few points are plotted to see if a trend develops.

For instance, the reactor delta T's were declining between 340 and 380 bbl/lb. There were no indications of high sulfur content, high chloride content, or out-of-the-ordinary feed properties to account for the loss in delta T and negative C_5^+ yield deviation trends of Fig. 5–12. The conclusion was that the catalyst was coked and needed regeneration. The recovery of delta T and C_5^+ yield after regeneration, together with the calculated coke burned, confirmed that the losses were due to coked catalyst. The plots show no reason for the next regeneration at 465 bbl/lb. The unit appeared to be performing well. The reformer was probably regenerated for plant maintenance and for scheduling of turnarounds on other units. Many reformer regenerations are carried out for plant convenience and not because of yield loss or high reactor inlet temperatures.

CHLORIDE AND WATER CONTROL

The most frequent reforming question is, "How do you control chloride and water?" A review of NPRA Q&A transcripts from 1974 through 1982 shows questions on both chloride and water control were discussed each year. The reason for such high interest was explored in chapter 4.

The proper chloride-and-water balance must be maintained for reforming catalyst to perform properly. If the chloride content of a modern bimetallic catalyst drops below 0.9 wt % or rises above 1.2 wt %, a change for the worse in catalyst activity or selectivity, or both, usually occurs.

The difficulty is that, at any given moment, the chloride content of the catalyst depends upon the partial pressures of hydrogen, water, hydrogen chloride, hydrocarbon, temperature, and probably carbon on the catalyst. A related difficulty is that the response to increasing or decreasing chloride injection is slow, sometimes requiring a few days to detect changes.

On Stream Catalyst Samplers

Recent developments give some relief to the problem of controlling chloride and water. With a continuous-regeneration reformer, circulating catalyst can be sampled at any time. Laboratory analysis establishes the amount of chloride on the catalyst, and adjustments can be made to chloride injection or water injection, as described later in this chapter. For those who do not have a continuous regeneration unit, the development of on-stream catalyst sampling devices permits obtaining samples for analysis without shutting down the unit. Fig. 5-15 is a schematic of a catalyst sampler. For units built before catalyst samplers were available, the sampler is installed through the catalyst dump nozzle in the bottom of the reactor.

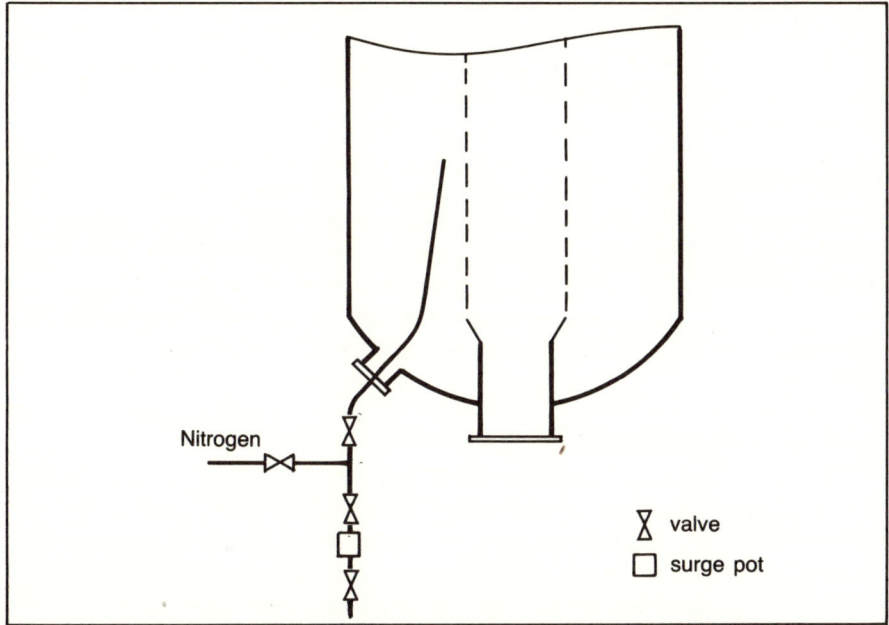

Fig. 5–15 Schematic diagram of on stream catalyst sampling device

The sampling device operates through a sequence of valves and nitrogen purges to remove a small quantity of catalyst. Although catalyst is sampled at only one point in the bed, the catalyst obtained generally gives a representative sample.[17]

Table 5–10 illustrates the use of catalyst samplers. The carbon and chloride content is shown after a regeneration, after 8 months on stream and after 16 months on stream. The buildup of carbon on catalyst in the fourth reactor is easily observed. This kind of information helps project the end of the cycle and aids planning the turnaround schedule.

TABLE 5-10
Analysis of catalyst obtained by on-stream sampling device

	Reactor No. 1		Reactor No. 4	
	Carbon, wt %	Chloride, wt %	Carbon, wt %	Chloride, wt %
After Regeneration[a]	0.87	0.87	0.4	0.10
8 Months[b]	2.10	0.98	12.1	0.77
16 Months[b]	2.60	0.93	16.2	0.72

[a]Samples obtained while unloading and screening
[b]Catalyst obtained through sampling device with unit in operation

Catalyst samplers in the first and last reactors are normally sufficient to establish catalyst condition.

Chloride Control

For those without continuous regeneration units or catalyst samplers, there are a few guidelines for determining catalyst condition. If there appears to be an increase in hydrocracking, indicated by a high yield of stabilizer overhead or by a low percentage of hydrogen in the recycle gas, there may be too much chloride on the catalyst. Stopping chloride injection for a few days may decrease hydrocracking.

Chloride in the recycle gas can also be a clue to weight percent chloride on the catalyst or if too much water is injected. Chloride in the recycle gas may be determined by a Draeger tube* or similar device. With bimetallic catalyst, 1–5 mol ppm of hydrogen chloride in the recycle is normal. Higher readings could mean excess chloride on the catalyst.

The trend in reforming control is to reduce both chloride and water in reactor charge as low as possible by hydrotreating and stripping. Adjustments are then made on the reformer by controlled chloride and water injections. Hydrotreating followed by stripping should reduce chloride in naphtha to 0.5 ppm wt.

There was a time, if a large amount of chloride on the catalyst was suspected, when the recommendation was to inject more water. If the chloride content was high, there was a surge of hydrocracking which was difficult to control until chloride had been reduced. Now the practice is to stop chloride injection until it becomes necessary to raise reactor

---

*A Draeger tube is a chemical in a glass tube which changes color in proportion to chloride concentration.

inlet temperature to maintain octane number. Then start injecting a small amount of chloride, 0.5–2 ppm wt.

The opposite of overchloriding is underchloriding, which can happen if chloride injection is not adequate to hold 0.9–1.2 wt % chloride on the catalyst. Underchloriding is deceptive. For a short time the unit appears to be operating exceptionally well. Reformate yields are good; separator gas yield is large, and hydrogen purity is high. But, eventually, temperatures must be raised to hold octane number. Coke laydown then increases, reformate yield decreases, and hydrogen purity in recycle gas decreases. The remedy for this situation is chloride injection, which may or may not succeed. If coke laydown is heavy, regeneration may be necessary. Chloride adjustment is needed most after regeneration, and the catalyst samplers' capability to sample and analyze catalyst during start-up justifies their cost.

Water Control

Water control must be used along with chloride control. As mentioned in chapter 4, alumina-base reforming catalysts require some moisture to activate the acid function designed into the catalyst. Moisture is almost universally controlled by measuring the moisture content of the recycle gas.

Bimetallic catalysts generally operate in the range of 10–20 mol ppm water in the recycle gas. Since reactor charge is usually stripped on the hydrotreater to 2–5 ppm wt water, a water injection of 3–4 ppm wt into the reformer feed will maintain 10–20 mol ppm water in the recycle gas.

There are two or three moisture analyzers on the market. Most reformers use these more or less successfully.[18] The key to success when using moisture analyzers is regular scheduled maintenance. Moisture analyzers are most necessary and make their greatest contribution to reforming during start-up after catalyst regeneration. Reactor temperatures must not be raised to make octane number until moisture is reduced at least below 50 mol ppm in recycle gas. Refiners generally rely on the moisture analyzer to protect the catalyst during start-up.

A unit which becomes too dry, that is, the water in the recycle gas falls below 5 mol ppm, usually shows an increase in methane yield. Because of very dry conditions in the reactor, chloride builds up on the catalyst, predominantly in the first reactor. The amount of hydrogen chloride in the recycle gas then becomes much less than usual. Once the lack of moisture is recognized, water injection should be started. This water must be injected cautiously to prevent hydrocracking.

TROUBLESHOOTING

The first signs of trouble on a cat reformer are loss of octane number, loss of reformate yield, loss of delta T, or low-purity hydrogen in the recycle gas. When trouble occurs, the first inclination is to blame the catalyst. The catalyst may not be performing as it should, but usually this is only a symptom of some other problem. A search should be made for the underlying cause.

LOSS OF OCTANE

A disturbing occurrence in reforming is finding that the octane number of the reformate has suddenly dropped one or two numbers, even though feedstock is still up to its usual quality. The first culprit, and one often overlooked, is a leak in the feed-to-reactor-effluent exchangers. Because the feed is always at a higher pressure than reactor effluent, the leak is feed into the effluent. Only a small percentage of low-octane-number feed (40–60 RON0) lowers the octane number of reformate.

A number of methods are used to detect exchanger leaks of this nature.[19,20] Some use tracers injected into the feed, some look for the presence in the reformate of easily dehydrogenated naphthenes such as methylcyclohexane, and some obtain the octane number of C_5^+ samples before and after the reactor effluent passes through the exchanger.

An underchlorided catalyst can cause loss of octane number because of a drop in catalyst activity. If catalyst samples cannot be taken, one way to indicate low chloride is raising reactor inlet temperatures and observing the temperature increase required to raise the octane of reformate by one number. If the increase is more than 5°F per RON0, underchloriding should be suspected. Increase chloride addition and observe the result.

Loss of octane number (catalyst activity) is also a symptom of nitrogen poisoning. The effects are deceiving because some aspects of the operation seem to be improving. For instance, hydrogen production increases, hydrogen purity goes up, and reactor delta T's increase. If there is ammonia in the recycle gas, look for the source of the nitrogen. Possibilities are the reactor charge, inadequate hydrotreating, poor efficiency in the HDS stripper, and use of a different filming amine inhibitor upstream of the reformer.

Sulfur's primary effect is on the loss of reformate yield, but high levels, say 10 ppm wt in the feed, also cause a loss of octane number. Too much sulfur in reformer feed is almost always a result of loss of hydrodesulfurization (HDS) catalyst activity or poor operation of the HDS unit stripper. Sulfur is one of the poisons most refiners run checks for on a regular schedule. Measuring the sulfur content of HDS raw naphtha charge, reformer reactor charge, and reformer recycle gas is usually part of the daily laboratory analyses.

The plant engineer can obtain hydrogen sulfide in recycle gas and stabilizer overhead gas at any time with a Draeger tube or similar device. If high sulfur content is detected, reactor inlet temperatures are lowered to 900°F until the source of contamination is eliminated. As mentioned in chapter 4, both sulfur and nitrogen are temporary poisons. When these are eliminated, the catalyst normally recovers.

Coking deactivates the catalyst, requiring higher reactor inlet temperatures to maintain the octane number. However, the deactivation normally takes place gradually over a long period. Reactor charge can be contaminated with high end-point material, in which case coking will be rapid and a noticeable loss in catalyst activity will occur in a few days.

Some sources of high end-point feed are fractionator upsets and exchanger leaks of gas oil or topped crude* on the crude unit. Seal oil on the recycle compressor has been known to leak into the recycle gas and enter the reformer reactors.

Another cause of loss of catalyst activity, and one which can have far more serious consequences than the loss of octane number, is the bypassing of the catalyst by the vapor in the reactors. This means less catalyst is in contact with reactor charge, or, in effect, higher space velocity through the remaining catalyst. Then a higher temperature is necessary to maintain octane number.

One cause of catalyst bypassing is buildup of fines or scale at a particular place in the catalyst bed. Fines or scale can also build up in the scallop annulus of a radial flow reactor. Catalyst bypassing can be due to the failure of reactor internals, which may allow catalyst to migrate out of the reactor.

The hazard of these buildup or failure conditions is that, at the point of the blocked (bypassed) catalyst, space velocities can be extremely low. Severe hydrocracking could then occur, even to the point of a temperature runaway. This phenomenon is discussed in the Safety section at the end of this chapter.

A redeeming feature of catalyst bypassing is that it generally reveals itself through increasing pressure drop. The pressures into and out of each reactor (the pressure differential across each reactor) are monitored by instrumentation with high-pressure alarms. Any increase in pressure is a signal of possible trouble.

Another reason for loss of octane number is catalyst poisoning by metals such as lead and arsenic. This phenomenon is discussed in this

---

*Reduced or topped crude is the residuum yielded from the crude unit.

chapter's Loss-of-Delta-T section because the most noticeable effect of metal poisoning is loss of delta T.

To summarize, the loss of reformate octane number is caused by one or more of the following:

- Change in feedstock quality
- Exchanger leak in feed/effluent exchanger
- Underchlorided catalyst
- Nitrogen in the feed
- High sulfur in the feed
- Coking of the catalyst
- Bypassing of catalyst
- Poisoning by metals.

LOSS OF YIELD

Some of the same troublemakers that cause loss of octane number can also cause loss of reformate yield. Those discussed fully in the previous section are only mentioned here.

Feedstock quality is directly related to reformate yield. As shown in Fig. 2–3, a lower N + 2A feed is expected to produce a lower yield of reformate of a given octane number.

An overchlorided catalyst causes a loss in yield because it promotes hydrocracking. The keys for detecting overchloriding are increased yield of stabilizer overhead gas, decreased yield of hydrogen, and a drop in hydrogen purity in the recycle gas. To correct an overchlorided catalyst, stop chloride injection and observe unit performance at least a week to ten days. Chloride comes off the catalyst slowly.

Sulfur contamination of reactor charge affects reformate yield to a greater extent than it affects octane number. Any increase in sulfur above 0.5 ppm wt is immediately reflected in loss of reformate yield and loss of delta T.

In fact, as noted in chapter 7, the sulfur guard catalyst was developed to reduce sulfur in reactor charge below that attainable with conventional hydrotreating. Sulfur guard catalyst is offered by some licensors of reforming processes.

Sulfur is easily and readily detected in recycle gas or stabilizer overhead gas by Draeger tube or similar instruments. Sulfur in the feed promotes coking of reforming catalyst. It also causes sintering* of platinum crystallites on the catalyst. Both sintering and coking degrade the catalyst activity and selectivity.

---

*Sintering is a phenomenon whereby reforming catalysts are susceptible to growth or agglomeration of metal crystallites.

PROCESS VARIABLES AND UNIT OPERATION   113

Poisoning of the catalyst by metals such as lead and arsenic causes a loss of reformate yield; this is discussed in the next section.

To recapitulate, loss of reformate yield may be caused by one or more of the following:

- Change in feedstock quality
- Overchlorided catalyst
- Sulfur in feed
- Coking of catalyst
- Poisoning by metals.

LOSS OF DELTA T

Loss of delta T across any or all reactors is usually immediately observable. Also, reforming technologists look at delta T as a qualitative indication of catalyst condition. Feedstock quality can affect delta T. The dehydrogenation of naphthenes is highly endothermic; a decrease in naphthene content of the feed will show a loss in delta T, particularly in the first reactor.

As previously mentioned, delta T can be affected by the H:HC ratio because the recycle gas functions as a heat sink. Therefore, an increase in the H:HC ratio results in lower delta T. As an illustration, a delta T of 180°F at an H:HC ratio of 5:1 and at 90% hydrogen purity recycle gas will be only 150°F at an H:HC ratio of 8:1 at the same 90% purity, or at 6:1 H:HC ratio and 96% purity.

Sulfur poisoning and coking of the catalyst were listed earlier as causes of reduction in delta T. Hydrocracking from an overchlorided catalyst also reduces delta T.

Metals Poisoning of Catalyst

Poisoning of the catalyst by metals such as arsenic and lead results in an alarming loss of delta T. This poisoning phenomenon is usually recognized by the change in delta T pattern through the reactors. The metal strongly adsorbs on the catalyst in the first reactor. It is here that loss of delta T is seen first. If the poisoning is allowed to continue, the dehydrogenation reactions move to the second reactor. Then the delta T in that reactor increases.

The typical pattern for metal poisoning is illustrated in Fig. 5–16. By the end of the run, the No. 1 and No. 2 reactor delta T's had crossed. When the catalyst was removed from the reactors, the catalyst from the No. 1 reactor contained 1,000 ppm arsenic. The same effect is experienced with other metals such as lead and copper.

The crude as received may be the source of heavy metals. Metals in the reformer charge can come from a treating unit or from contami-

114 CATALYTIC REFORMING

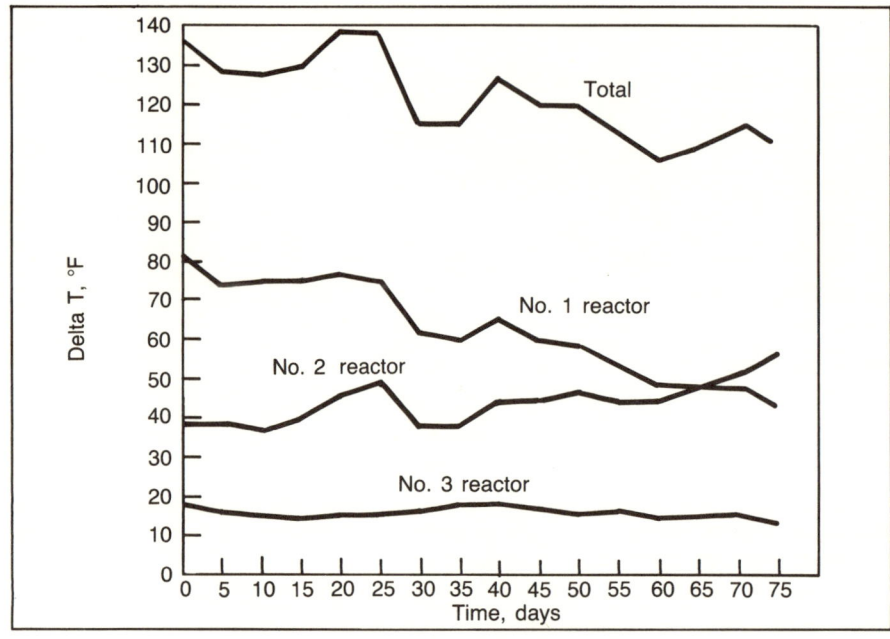

Fig. 5–16 Delta T pattern for arsenic poisoning

nation of reformer feed from reprocessing off-specification products. The rerunning of off-specification leaded gasoline on crude units has often caused lead poisoning of catalyst on reforming units.

In response to a question at the 1975 NPRA about distribution and decomposition of tetraethyl lead (TEL) in crude units Unzelman reported that, when slop is added to the crude mix, as much as 5% of the lead in the slop can appear in the gasoline fraction (90–400°F).[21] He further reported that reducing the end point from 400°F to 365°F reduces the lead in the gasoline cut by a factor of from two to three. In the case of tetramethyl lead (TML), Unzelman estimated perhaps 75% of the lead can appear in the gasoline fraction. The thermal stability of TML is greater than that of TEL. The boiling point of TML is 230°F, compared to 390°F for TEL.

A study some years ago indicated that processing a tetraethyl-lead-contaminated crude distributed lead throughout all fractions, from light gasoline to topped crude.[22] Most of the lead was found in the residual oil, but enough was found in the reformer naphtha stream to severely poison reforming catalyst.

Hydrotreaters that prepare feed ahead of reformers use cobalt-moly or nickel-moly on an alumina catalyst. Arsenic and lead are adsorbed from naphtha (HDS feed) at the cost of poisoning the HDS catalyst.

Table 5–11 shows analysis of arsenic and lead in raw naphtha going to the HDS unit and to the reformer feed after the HDS treating. Metals such as arsenic and lead are adsorbed chromatographically by HDS catalyst. Analysis of samples obtained when dumping cobalt-moly catalyst from an HDS reactor are shown in Table 5–12. Some refiners rerun off-specification leaded gasoline on the cat-cracker fractionator.

TABLE 5–11
Removal of arsenic and lead from reformer naphtha feed with cobalt-moly catalyst

	January 1971	July 1971	April 1972	February 1973	March 1974	December 1983
Arsenic In, ppb wt	3	3	6	10	1.3	34.9
Out, ppb wt	0	1	0	<1	1.0	<0.6
Lead In, ppb wt	<5	<5	0	<5	21	<2.7
Out, ppb wt	<5	<5	0	<5	3	<2.7
Sulfur In, ppm wt						140
Out, ppm wt						0.1

TABLE 5–12
Arsenic and lead content of used cobalt-moly HDS catalyst

Reactor Sample	Arsenic, wt %	Lead, ppm wt
Top	0.49	297
Middle	0.26	111
Bottom	0.15	49

To summarize, loss of delta T may be caused by one or more of the following:

- Reduced naphthene content of reactor charge
- Increase in H:HC molar ratio
- Sulfur poisoning of catalyst
- Coking of catalyst
- Hydrocracking (overchlorided or coked catalyst)
- Metal poisoning of catalyst by arsenic, lead, copper, or similar heavy metals.

REFORMATE COLOR

Reformate is normally water-white in color, but, as reforming severity increases and as reformate octane number increases, reformate may

116 CATALYTIC REFORMING

become colored.[23] The color ranges from pale yellowish straw to bright yellow.

The severity at which the reformate takes on color varies from unit to unit. Some report color as low as 95–97 RON0; others do not see it until reformate octane number clear rises above 100. The consensus is that the color is a result of a polynuclear aromatic, coronene. Coronene has also shown up as red deposits in the reformate stabilizer reboiler and in one unit as a viscous red liquid, probably washed from exchangers after regeneration.[24]

FOULING AND CORROSION

Catalytic reformers by their very nature are subject to fouling and corrosion of equipment. The presence of chlorides, ammonia, sulfur, hydrogen, and moisture—at the right temperature and pressure conditions—can lay down fouling deposits or be corrosive, or both. Nearly as many questions are asked at NPRA Q&A sessions on this subject as on chloride and water control.[25]

The fouling of exchangers, lines, compressors, and fractionators is attributed primarily to deposits of ammonium chloride, iron sulfides, ammonium sulfides, and polymers.

Chlorides

Even though nitrogen in the feed is below specified limits, traces of nitrogen, over lengthy periods, will deposit ammonium chloride in the equipment. Sometimes the deposits are effectively removed by water washing lines, exchangers, and fractionators, without shutdown. However, water wash of compressor suction valves or rotors is not recommended without a shutdown. Generally, mechanical cleaning is preferred over a water wash.

Some operators report successful removal of ammonium chloride by raising temperature of vapors-to-compressor-suction above 400°F. Ammonium chloride sublimes* above 400°F.

One precaution about which there is no controversy is that a centrifugal compressor should not be opened for inspection without first washing with a neutralizing solution of dilute sodium carbonate. Otherwise, the machine can be severely damaged by stress corrosion cracking. Reformate fractionators become fouled with ammonium chloride or ammonium sulfides. These fractionators are usually freed of such deposits by water washing without opening up the vessel.

---

*Passes from solid to vapor without going through a liquid state

Polymers

Polymer formation was a serious problem on some of the first reforming units. Polymers deposited in heat exchangers and heater tubes greatly reduced heat transfer rates. The polymer deposits are attributed to oxygen from entrained air (air trapped in the liquid) in the feed naphtha. The nitrogen blanketing of naphtha feed storage tanks and direct feed to the reformer without intermediate storage provides satisfactory control of polymer deposits.

On some units, oxygenated hydrocarbons present in the feed also cause polymer formation. Hydrotreating with a catalyst high enough in activity to convert the oxygenated compounds to water and hydrocarbons eliminates polymers.

One other method for control of fouling is the use of *inhibitors*. Usually, injection of a few parts per million of a filming amine is effective. Vendors of process chemicals offer a number of amine inhibitors, each designed for application under specific operating conditions.

Corrosion

Although fouling deposits cause pressure drop and heat transfer troubles, *corrosion* is a more serious problem. Corrosion can damage equipment and, if undetected, can result in sudden and disastrous rupture of high-pressure lines, exchangers, or vessels.

In catalytic reformers, corrosion is often traced to chloride or sulfur, or both. Chloride corrosion is most serious at the point where water condensation begins and hydrogen chloride is present. The most widely used chloride corrosion control method is neutralization with aqua ammonia, either alone or in combination with a filming amine. Sometimes a filming amine is used by itself. The most expensive and least-used method is replacement of equipment with alloys that resist hydrogen chloride corrosion.

Some refiners maintain a rigid inspection schedule and replace equipment when metal thickness reaches a specified minimum. Sulfur corrosion in reformers is attributed to hydrogen sulfide, most often found in the overhead condenser and in the accumulator of the reformate fractionator. Control in those locations is generally by application of a filming amine.

REFORMING CRACKED NAPHTHAS

Reforming of cracked naphtha does not differ from reforming virgin naphthas, provided contaminants are removed to levels acceptable for reforming feedstock. Cat-cracked naphtha and coker naphtha are products from pyrolysis of the heavier fractions of crude oil. These naphthas have a high content of sulfur, nitrogen, and olefins (Table 5–4). Pyrolysis naphthas (from cracking ethane, propane, butane, naphtha, or distillate) may not contain as large a percentage of sulfur and nitrogen as the cat-cracked and coker naphthas. However, pyrolysis naphthas contain more monoolefins and diolefins than the cat-cracked and coker naphthas.

The point in reforming cracked naphtha is to raise octane number. Coker gasoline is seldom above 88 RON0 and is usually in the 75–85 RON0 range. Cat-cracked gasoline is as high as 96 RON0 with good feedstocks, but more and more residual stocks are being included in cat-cracker feed. So the cat-cracked gasoline is more often in the range of 87–92 RON0.

Pyrolysis gasoline is rich in aromatics and, after hydrotreating to remove diolefins, has a research octane number clear in the range of 94–107, sufficient to blend to gasoline or, alternatively, to utilize as excellent charge to a BTX reformer.

A number of refiners are reforming both cat-cracked and coker naphthas.[26,27] Once the contaminants are reduced to acceptable levels, cracked naphthas reform as any other feedstocks, according to N + 2A. Hydrotreated cracked naphthas are rich in aromatics and naphthenes (high N + 2A) and give large yields of high-octane-number reformate.

The primary concern of refiners is not whether cat-cracked naphtha can be reformed but is the selection of the proper-boiling-range feed. Unlike straight-run naphtha, certain fractions of cat-cracked naphtha have relatively high octane numbers.

An illustration is Table 5–13 (pp. 120-121), which shows the octane numbers of 10% fractions of a cat-cracked gasoline. These data are graphed in Fig. 5–17. The low-boiling-fraction IBP to 180°F, repre-

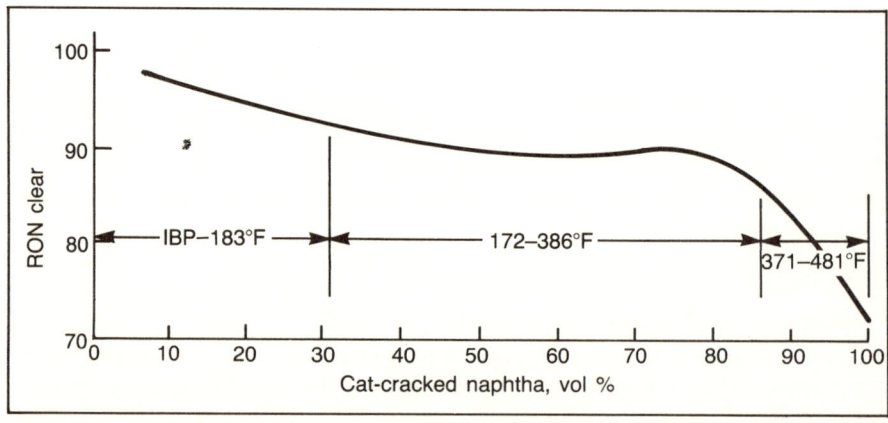

Fig. 5–17 Fractionation of cat-cracked naphtha

senting about 30% of the total gasoline, has a reasonably high research octane number clear, in the range of 92–98, and is suitable for gasoline blending. The heavy fraction, 370–480°F, representing about 15% of the total gasoline, has a low research octane number clear, in the range of 70–85.

As discussed previously, the high-boiling hydrocarbons coke rapidly in the reformer and shorten catalyst cycle life. Therefore, the logical feed for catalytic reforming from this stock is the 170–380°F center cut, representing about 55 liquid vol % of the total cat-cracked gasoline.

A similar conclusion may be drawn from the cat-cracked gasolines of Fig. 5–18.[28] These two gasolines represent low, medium, and high conversions on the cat cracker. These gasolines show a low-boiling (IBP = 180°F) fraction of high octane number, a low-octane center cut (180–270°F), and a high-octane heavy end (270–430°F). The conclusion is that the 180–270°F fraction is the choice for reforming feedstock.

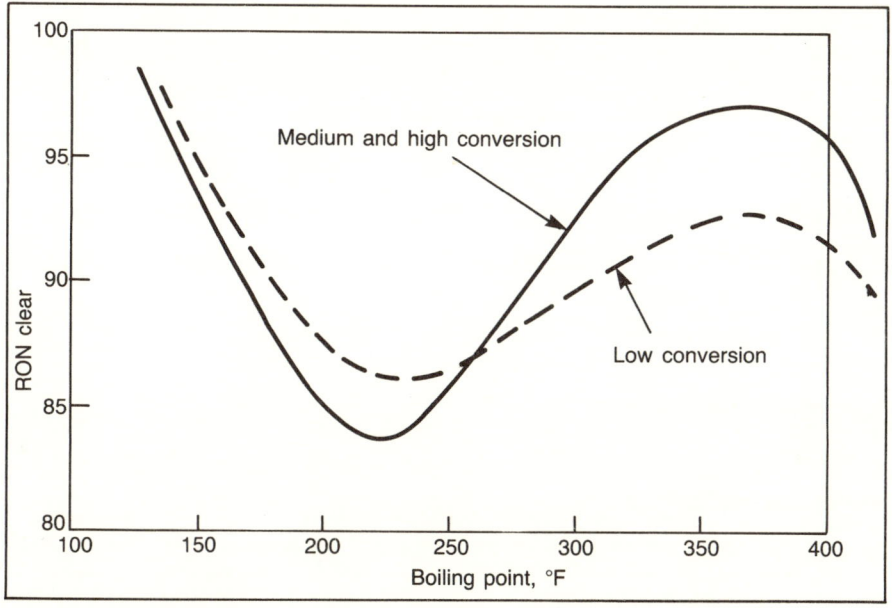

Fig. 5–18 RON clear vs boiling point for cat-cracked naphtha *(after Gladrow, et al., ref. 28)*

Sulfur is a reforming feedstock contaminant common to cracked naphthas and often in a rather high concentration. For example, the sulfur content of heavy cat-cracked naphtha (250–430°F) from a number of cat crackers charging a wide variety of feedstocks ranged from 2,400 to 6,500 ppm wt.[29]

TABLE 5-13
Octane numbers of cat-cracked naphtha fractions

	Feed 7	7A	7B	7C	7D	7E
OPERATING CONDITIONS						
Reflux Ratio	—	5:1	5:1	5:1	5:1	5:1
Top Column Temperature, °F	(81 IBP)	95	142	169	203	234
Yield, Vol %	100.0	10.15	10.18	10.17	10.12	10.11
PROPERTIES						
API Gravity, 60°F	53.9°	87.7°	79.5°	70.1°	63.5°	55.6°
Specific Gravity, 60:60°F[a]	0.7632	0.6455	0.6706	0.6984	0.7256	0.7563
ASTM D 86 Distillation[b]						
IBP, °F	111	85	104	143	172	206
5%, °F	135	88	111	147	179	208
10%, °F	149	89	114	149	180	210
20%, °F	168	90	115	149	181	210
30%, °F	191	91	117	150	182	211
40%, °F	217	91	119	151	184	214
50%, °F	245	92	121	153	186	216
60%, °F	275	93	123	154	188	217
70%, °F	309	93	127	155	190	219
80%, °F	345	94	133	157	193	222
90%, °F	383	97	140	160	197	227
95%, °F	411	99	145	163	202	232
EP, °F	440	103	150	183	210	245
Recovery, Vol %	98.5	98.0	98.0	98.0	98.0	99.0
Residue, Vol %	1.5	1.0	1.0	1.0	1.0	1.0
Loss Vol %	0.0	1.0	1.0	1.0	1.0	0.0
Research Octane Number						
+0.0 TEL	89.6	96.5	95.0	92.9	90.2	90.0
+1.0 g TEL/gal	93.7	100.5	99.3	97.5	95.4	95.6
Motor Octane Number						
+0.0 TEL	78.2	82.1	80.6	78.7	77.7	76.6
+1.0 g TEL/gal	81.3	85.9	84.4	82.8	82.3	81.7

[a] 60:60°F is liquid at 60°F : water at 60°F
[b] Evaporated temperature, corrected to 760 mm mercury

Pilot unit results from hydrotreating of a cat-cracked naphtha are shown in Table 5–14. Note that both the sulfur and nitrogen content of the 167–420°F fraction are acceptable for reforming with bimetallic catalyst. The olefins in the feed are hydrogenated to paraffins and naphthenes.

	7F	7G	7H	7J	7K	7L	7M	7N
	5:1	5:1	5:1	5:1	5:1	5:1	5:1	5:1
	274	314	332	355	382	407	436	436+
	10.09	10.08	5.03	5.09	5.03	5.03	5.05	3.87
	50.8°	44.7°	41.4°	39.4°	37.5°	34.8°	32.7°	27.5°
	0.7762	0.8031	0.8184	0.8280	0.8373	0.8509	0.8618	0.8899
	236	272	301	321	347	371	397	431
	242	277	308	329	351	376	402	435
	244	279	309	330	352	378	403	436
	246	281	311	331	353	379	404	437
	248	282	313	332	355	380	405	438
	249	284	315	333	357	381	406	439
	251	286	316	335	359	383	408	440
	253	287	318	336	360	385	410	441
	256	290	320	338	363	386	411	443
	259	293	323	341	366	388	413	445
	265	298	327	346	371	392	417	447
	271	307	331	351	376	397	421	458
	280	321	338	362	386	405	439	481
	99.0	99.0	99.0	99.0	99.0	98.5	99.0	99.0
	1.0	1.0	1.0	1.0	1.0	1.0	1.0	0.5
	0.0	0.0	0.0	0.0	0.0	0.5	0.0	0.5
	88.9	89.9	90.2	89.3	86.2	83.9	76.0	71.2
	94.0	94.1	94.0	93.0	90.5	88.1	81.8	75.2
	78.0	79.2	79.3	79.0	76.4	73.9	67.2	61.8
	81.6	81.9	82.3	82.0	80.1	77.4	74.6	64.0

Due to the severity of hydrotreating, some of the aromatics are also hydrogenated. Saturation of olefins and aromatics consumes 644 scf of hydrogen per barrel of feed. Much of the hydrogen is recovered upon reforming, plus hydrogen from naphthenes initially in the naphtha. There is a net yield of hydrogen from the combined HDS/reforming operation.

TABLE 5–14
Hydrotreating of cat-cracked naphtha (pilot unit)—nickel-moly HDS catalyst

	Feed	Total Liquid Product	167–420°F Fraction of Product
Yield, Vol % Feed	100.0	—	59.8
API Gravity	56.0°	58.7°	50.8°
Sulfur, ppm wt	840.0	1.1	0.3
Nitrogen, ppm wt	39.0	0.3	0.3
RON0	90.5	—	^{a}70
RON + 3 ml	97.3	—	^{a}87
Bromine Number	66.0	0.3	0.4
Paraffins, Vol %	30.3	52.3	37.3
Olefins	31.1	0.9	1.3
Naphthenes	10.1	23.4	30.5
Aromatics	28.5	23.4	30.9
N + 2A	67.1	70.2	92.3
H_2 Consumed, scf/bbl	644	—	—
Distillation, ASTM D–86, °F			
IBP	109	109	190
10 vol %	139	145	224
30 vol %	—	183	248
50 vol %	230	237	280
70 vol %	—	302	321
90 vol %	392	385	370
EP	435	439	401

^{a}Estimated

Another observation is the loss of octane number in hydrotreating because of saturation of olefins and aromatics. A compensating factor is the high naphthene and aromatic content, 92.3 (N + 2A), of the 167–420°F fraction. This N + 2A indicates a relatively large reformate yield, even when reforming to 100+ RON0.

A second example of hydrotreating and reforming cat-cracked naphtha is outlined in Table 5–15. In this pilot test, the cat-cracked naphtha was produced from cat cracking gas oil. In this instance, the HDS product was reformed to 100 RON0 and 104 RON0 at high yields because the N + 2A of the naphtha after hydrotreating was 105.

Pilot unit tests on reforming fractions of heavy cat-cracked gasoline without hydrotreating reported results similar to the HDS/reforming route (see Tables 5–16, 5–17, and 5–18).[30] The reformates ranged from 99.7 to 104.5 RON0. As expected, the aromatic content of such high-octane reformates was in the 70–85 liquid vol % range. Note that the contribution of reforming was a net aromatic production of 18.1–26.9 liquid vol % of charge.

TABLE 5-15
Hydrotreating and reforming cat-cracked naphtha (pilot unit)

	Raw Naphtha HDS Charge	HDS Product Reformer Charge	Reforming Products	
API Gravity	37.1°	41.3°	36.9°	34.1°
Sulfur, wt %	0.6	0.01	—	—
Bromine Number	58	0.2	1.0	1.2
Total Nitrogen, ppm wt	800	1.5	—	—
Basic Nitrogen, ppm wt	711	0.1	NIL	NIL
Paraffins, Vol %	[a]19	22	20	[b]19
Olefins	20	0	1	0
Naphthenes	14	31	6	—
Aromatics	47	47	73	81
Distillation, ASTM D–86, °F				
IBP	252	226	175	164
10%	279	259	246	240
30%	302	293	282	284
50%	326	317	308	306
70%	353	347	338	334
90%	379	372	384	386
EP	400	416	465	464
H_2 Consumed, scf/bbl	—	480	—	—
Yields				
Hydrogen, scf/bbl	—	—	501	499
Methane & Ethane, scf/bbl	—	—	97.7	182.4
Propane, Liquid Vol %	—	—	3.9	5.8
Butanes, Liquid Vol %	—	—	4.1	6.0
C_5^+ Liquid Vol %	100.0	102.7	89.8	84.6
RON0	95.6	76.9	100.0	104.0
RON + 3 ml	97.9	89.2	103.8	107.5
MON0	82.1	70.0	90.0	93.4
MON0 + 3 ml	84.4	81.0	93.3	97.1

[a]Estimated PONA
[b]Paraffins + naphthenes

TABLE 5-16
Reforming cat-cracked gasoline[a]

Properties of Cat-cracked Gasoline Fractions

Cut Number	Total	1	2	3	4	5
Fraction Boiling Range, °F[b]	—	235–270	270–300	300–340	340–412	235–412
Fraction Yield, Vol % of Total	100	10.9	8.7	8.2	14.3	42.1
Sulfur, wt %	0.018	0.020	0.017	0.018	0.025	0.021
Octane Number						
F-1 clear[c]	86.7	86.4	92.1	95.0	84.5	88.7
F-1 + 3 cc TEL	95.9	—	—	—	—	—
Gravity, °API	55.9	47.6	41.6	38.1	36.1	40.6
ASTM Distillation, °F						
5%	130	249	272	300	354	271
10%	142	251	273	303	354	277
30%	180	255	282	308	360	294
50%	238	259	285	312	365	308
70%	300	264	290	317	372	328
90%	366	273	300	324	384	364
EP	406	293	342	365	418	394
Chemical Composition, Vol %						
Paraffins	42.2	29.0	22.2	19.1	22.8	23.6
Monocyclic Naphthenes	15.4	21.4	14.7	10.1	12.3	14.7
Dicyclic Naphthenes	1.4	0.3	1.3	2.1	6.7	3.0
Noncyclic Olefins	9.8	7.6	4.6	16.6	—	6.1
Cyclic Olefins	2.0	3.3	1.8	0.2	—	1.3
Monocyclic Aromatics						
C_6	0.9	0.3	0.3	—	—	0.1
C_7	3.7	8.1	0.7	—	—	2.3
C_8	9.7	28.7	38.4	12.3	—	17.8
C_9	9.7	1.2	15.5	37.2	15.5	16.0
C_{10}	0.4	—	0.3	1.6	22.8	8.1
C_{11}	1.2	—	—	—	5.7	2.0
C_{12}	0.3	—	—	—	0.8	0.3
Indans	2.5	—	0.2	0.8	9.8	3.4

Naphthalenes
C_{10} 0.5   —   —   —   3.0   1.0
C_{11} 0.1   —   —   —   0.6   0.2

[a] Courtesy American Institute of Chemical Engineers; after Kress, Nagel, and Reif, ref. 30
[b] Initial to 235°F cut was not reformed
[c] F-1 = Research octane number ASTM designation: D908 for values less than 100, D1656 for values over 100

TABLE 5-17
Reforming cat-cracked gasoline

Reformer Operating Conditions

Pressure, psig	500
LHSV,[a] hr^{-1}	3.0
Recycle Ratio (Moles Total Gas)	10:1
Reactor Inlet Temperature, °F	925 and 950

Reforming Data and Product Properties

Fraction Boiling Range, °F	235–270	270–300	300–340	340–412	235–412	235–412
Reactor Inlet Temperature, °F	925	925	925	925	925	950
Product Yields, Vol % Charge						
C_5^+	86.6	90.8	92.0	90.8	91.9	89.1
nC_4	3.5	2.3	1.9	1.9	1.9	2.5
iC_4	2.9	1.7	1.2	1.1	1.3	1.7
C_3	4.7	3.5	3.2	3.6	3.0	4.3
C_2 and Lighter, Fuel Oil Equivalent	4.1	3.4	3.1	4.3	3.2	4.2
HPSG,[b] scf/bbl	488	330	237	404	330	381
Hydrogen Purity, HPSG vol %	84.7	81.8	78.0	78.4	85.3	80.0
LPSG,[c] scf/bbl	96	89	93	93	83	114

[a] LHSV = Liquid hourly space velocity (vol liquid feed/hr/vol catalyst)
[b] HPSG = High-pressure separator gas
[c] LPSG = Low-pressure separator gas
[d] F-2 = Motor octane number ASTM designation D357
[e] Reformed at 950°F

TABLE 5-17
Reforming cat-cracked gasoline—cont'd

Reformer Operating Conditions

C_5^+ Reformate Octane Numbers						
F-1 clear	99.7	103.5	104.5	101.7	101.0	102.8
F-1 + 3 cc TEL	104.2	106.8	107.4	103.6	104.6	106.0
F-2 clear[a]	89.3	93.0	94.4	90.8	90.0	92.2
F-2 + 3 cc TEL	94.3	97.3	98.8	93.7	94.4	97.0
Sensitivity (+3 cc TEL)	9.9	9.5	8.6	9.9	10.2	9.0
Octane Number (F-1 Clear) Increase	+13.3	+11.4	+9.5	+17.2	+12.3	+14.1

C_5^+ Reformate Properties

Fraction Boiling Range, °F	235–270	270–300	300–340	340–412	235–412	235–412
Gravity, °API	41.7	38.0	35.2	32.3	36.9	35.8
ASTM Distillation, °F						
5%	188	201	238	204	235	228
10%	216	238	276	212	252	232
30%	256	276	304	312	282	258
50%	268	288	312	—	300	292
70%	277	296	320	—	320	320
90%	288	320	334	—	358	358
EP	340	359	406	—	450	456
Reid vapor pressure, psig	2.8	2.3	1.7	1.5	1.3	2.2

TABLE 5-18
Chemical composition of C_5^+ reformates

Fraction Boiling Range, °F	235–270	270–300	300–340	340–412	235–412	[a]235–412
Chemical Composition, Vol %						
Paraffins	26.6	14.5	13.0	13.5	22.2	16.8
Monocyclic Naphthenes	2.0	1.5	0.5	1.0	2.0	1.1
Dicyclic Naphthenes	0.1	—	0.1	0.1	0.2	0.1
Noncyclic Olefins	0.2	0.5	0.8	—	0.2	—
Monocyclic Aromatics						
C_6	0.5	0.5	0.4	0.4	0.4	0.5
C_7	11.9	3.4	1.8	0.9	3.6	4.6
C_8	50.0	50.1	21.8	5.4	26.9	29.8
C_9	8.1	26.2	52.1	23.5	24.4	26.4
C_{10}	0.5	2.7	7.3	26.2	10.4	10.8
C_{11}	—	—	0.4	7.9	2.7	2.5
C_{12}	—	—	—	2.2	0.7	0.6
C_{13}	—	—	—	0.2	—	—
Indans	0.1	0.5	1.5	10.2	3.8	3.6
Naphthalenes						
C_{10}	—	0.1	0.2	3.3	1.0	1.3
C_{11}	—	—	0.1	3.1	0.9	1.3
C_{12}	—	—	—	2.1	0.6	0.7
Total Aromatics	71.1	83.5	85.6	85.4	75.4	82.1

Aromatics Produced by Reforming

Fraction Boiling Range, °F	235–270	270–300	300–340	340–412	235–412	[b]235–412
Nonaromatics produced, bbl/100 bbl Charged						
C_7	2.3	2.5	2.0	1.2	1.3	2.1
C_8	14.6	7.1	7.8	4.9	6.9	8.7
C_9	5.8	8.3	10.8	5.9	6.4	7.5
C_{10}	0.4	2.1	5.1	1.0	1.5	1.5
C_{11}[c]	—	0.3	1.2	6.4	2.0	2.0
Total	23.1	20.3	26.9	19.4	18.1	21.8

[a] Reformed at 950°F
[b] Reformed at 950°F
[c] Includes monocyclic aromatics, indans, and naphthalenes

One other reforming feedstock is hydrocracker naphtha, or *hydrocrackate*. The low-boiling fraction of total hydrocrackate, like light cat-cracked naphtha, is of high enough research octane number clear, 90–92, to not be reformed.

The heavy hydrocrackate is rich in naphthenes and aromatics but of low octane number. This hydrocrackate is then reformed to raise the octane. The properties of a heavy hydrocrackate are listed in Table 5–19. The octane numbers are too low for gasoline blending, but the N + 2A of 98 liquid vol % makes the heavy hydrocrackate a premium feedstock for reforming. This hydrocrackate was reformed in a pilot unit to 100 RON0 to yield 86.3 liquid vol % C_5^+ reformate. Hydrogen yield was 977 scf/bbl.

TABLE 5–19
Properties of a heavy hydrocrackate

Gravity, °API	46.1
Sulfur, wt %	nil
Bromine Number	0.3
Basic Nitrogen, ppm wt	0.4
Paraffins, wt %	25
Olefins	0
Naphthenes	52
Aromatics	23
RON0	69.9
RON + 3 ml	86.0
MON0	—
MON + 3 ml	81.5
Distillation, ASTM D–86, °F	
IBP	195
5%	219
10%	228
30%	259
50%	296
70%	337
90%	377
EP	406

Hydrocrackate, having passed through a severe hydrotreating environment in the hydrocracker unit, is sufficiently free of contaminants to be charged directly to a catalytic reformer without hydrotreating.

REGENERATION

Regeneration of spent catalyst is as much a part of catalytic reforming operations in the 1980s as chloride/water control. At one time, regeneration meant only burning carbon off the catalyst. Regeneration now means a sequence of procedures, and carbon burn is only one of the steps. To a reforming technologist, regeneration includes a proof burn, rejuvenation to redistribute metal crystallites, reduction, chloriding, and sulfiding. If the catalyst has been deactivated by sulfur, there is even a sulfur removal procedure.[31] Each catalyst manufacturer has a regeneration procedure developed for its proprietary catalyst. There are variations, but the general procedure seems to be about the same for bimetallic or multimetallic catalysts. The order of events is:

- Shutdown and purge
- Carbon burn
- Oxidation or proof burn
- Rejuvenation
- Reduction
- Chloriding
- Sulfiding.

For reasons of confidentiality, regeneration procedure cannot be discussed in detail. However, some items of interest can be reviewed.[32,33]

SHUTDOWN AND PURGE

Shutdown of a catalytic reformer calls for a gradual reduction of the temperature of the catalyst to the 800–850°F range. Reactor charge is then withdrawn slowly, but recycle gas circulation is continued until catalyst temperature is in the 700–750°F range. If catalyst is to be removed from the reactors, then the carbon burn should be done before the removal.

If the catalyst is to be shipped off for ex-situ or merchant regeneration, then it should be cooled to 100–150°F before unloading. The carbon on the catalyst is pyrophorric and burns on contact with air. The procedure is to unload the catalyst directly into the shipping drums (which have been flushed with nitrogen) and immediately exclude air by covering with the drum lid.

For in-situ regeneration, the catalyst is screened and reloaded into the reactors for the oxidation-proof burn step. In preparation for the carbon burn, the system is evacuated and purged with nitrogen.

CARBON BURN

The carbon burn is usually the most time-consuming part of a regeneration. The burning or oxidation of carbon off the catalyst must be strictly controlled to avoid excessive temperatures which damage the catalyst. The temperature of regeneration gas flowing out of any reactor is customarily limited to an 850°F maximum. The need for caution during a carbon burn must be emphasized. Some refiners have learned to

130 CATALYTIC REFORMING

their dismay that it is not wise to raise the oxygen percentage to speed up burning of coke. If the oxygen content is increased, temperature rises rapidly. There is no way to reduce the temperature until the high-temperature front has passed through the reactor.

An illustration of calculated temperature rise in regeneration of catalyst by oxidation of carbon to carbon dioxide using air and nitrogen diluent is Fig. 5–19. This chart was constructed using the heat of reaction for $C + O_2 \rightarrow CO_2$ and the specific heats of air, nitrogen, carbon dioxide,

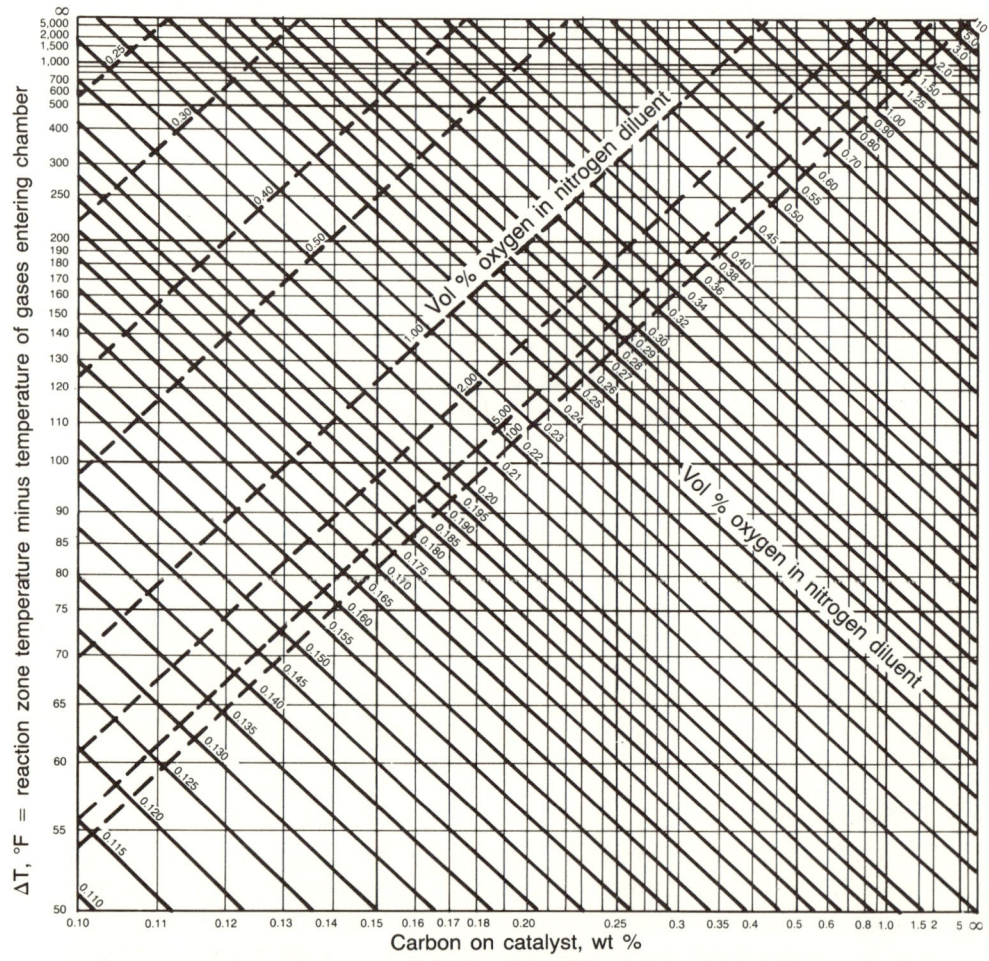

Fig. 5–19 Regeneration of catalyst by oxidation of carbon to carbon dioxide, using air and nitrogen diluent

and the catalyst. Although calculated using certain premises, the chart demonstrates why temperatures rise so much with increased volume percent oxygen when carbon is 1 wt % or more of the catalyst.

For example, if the carbon on the catalyst is 1.0 wt % and the volume percent oxygen in the regeneration gas is 1.0 vol %, the reaction zone temperature is 420°F higher than that of the gases entering the reactor. Fig. 5–19 also illustrates the reason for the recommendation that the carbon-burn step be carried out with the regeneration gas containing a maximum of 0.2–0.5 vol % oxygen. In fact, to initiate the burn, air or oxygen is admitted very slowly, until the burn is established.

A bimetallic catalyst unloaded from a commercial unit before carbon burning contained 4.3–16.9 wt % carbon. These figures are low for carbon on bimetallic catalyst. Carbon contents of 20–30 wt % and higher have been reported. Refiners are generally comfortable if the temperature rise across a reactor is in the range of 50–100°F. Therefore, reactor inlet temperatures are held at 700–750°F and reactor outlet temperatures at an 850°F maximum.

Temperature rise during regeneration on a commercial unit is shown in Fig. 5–20. This is a parallel-burn regeneration. To shorten carbon-burn time, oxygen is admitted to the inlets of both the first and fourth reactors. Note that temperatures began to rise in No. 1 and No. 4 reactors at about the same time. Reactor No. 2 did not start burning until oxygen broke through No. 1 reactor. Reactor No. 3 did not start burning until oxygen broke through No. 2 reactor. The entire carbon burn required about 36 hr for nearly 100,000 lb of catalyst.

Sometimes a reactor delta T does not return to near zero but continues to indicate an exotherm long after the peak temperature has passed. On one unit, the first three reactors completed burning in 35 hr, but the fourth reactor was still showing a 30°F exotherm after 84 hr. The long, trailing exotherms are caused by very high carbon on the catalyst, regeneration gas bypassing part of the catalyst, or hydrocarbons left in low spots during initial purging. The only viable option is to stop the carbon burn, unload the reactor, screen the catalyst, and finish the oxidation after reloading.

Current practice with bimetallic catalyst regeneration is to inject chloride into the No. 1 reactor inlet and sometimes also into the No. 4 reactor inlet during the carbon burn.[34] To prevent corrosion downstream of the reactors, a dilute solution of sodium carbonate or sodium hydroxide is circulated from the product separator to the inlet of the reactor effluent/feed exchangers. This neutralizing solution is maintained at 6–7 pH by addition of fresh caustic.[35] The amount of chloride injected is proprietary with each catalyst supplier.

132 CATALYTIC REFORMING

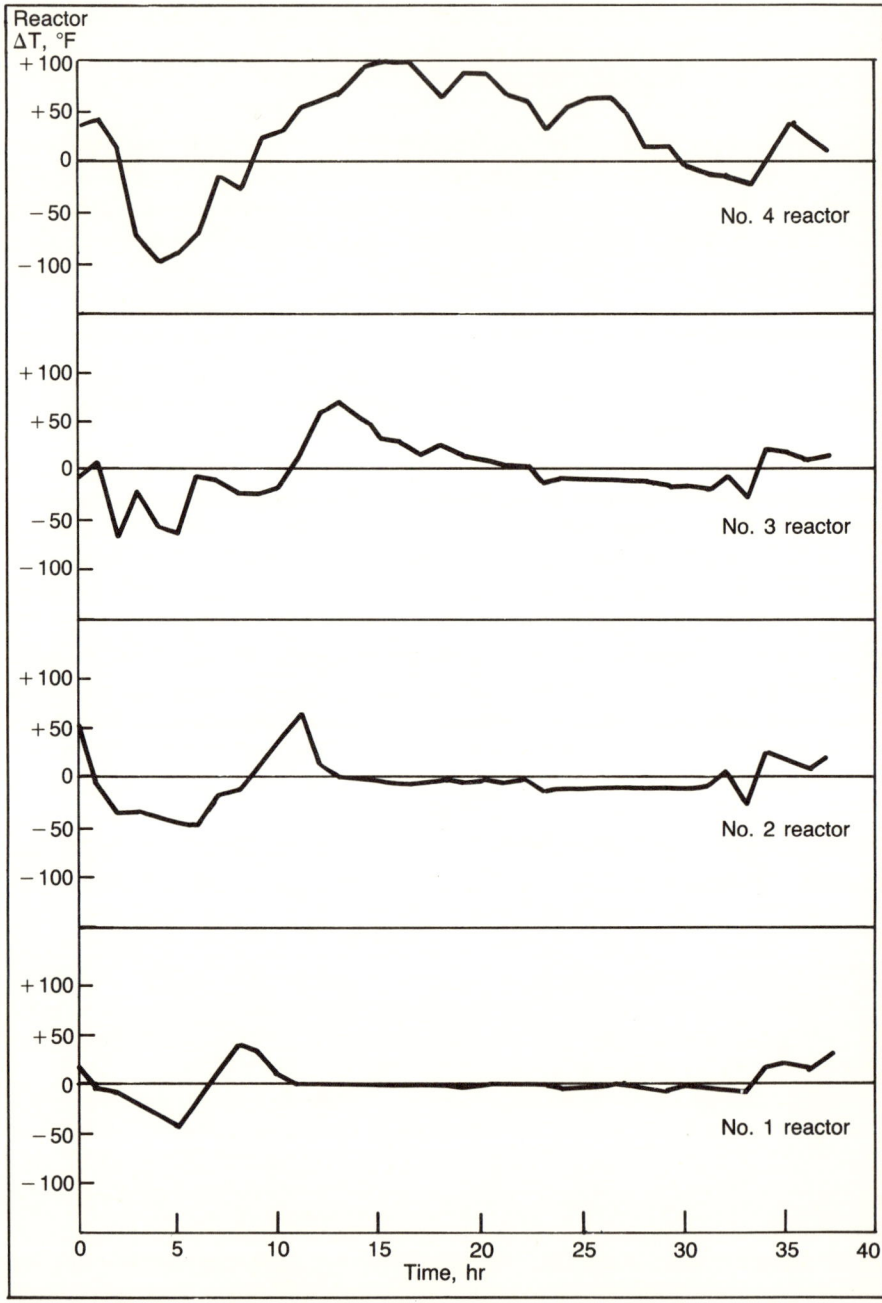

Fig. 5–20 Reactor delta T's during carbon burn

As reforming severity increases and regenerations are more frequent, efforts have been made to shorten downtime (refiners need hydrogen) by increasing the rate of carbon burning. Reactor pressure during carbon burn has been raised from the 70–100 psig range to the 400–425 psig range, which increases the pounds of oxygen per hour but not the percent oxygen in the regeneration gas. Refiners report 40–50% reduction in regeneration time.[36,37] The pressure is increased by utilizing auxiliary compressors and by replacing air with liquid oxygen.[38,39]

PROOF BURN

After completion of the carbon burn, the catalyst may be cooled down and unloaded for screening or for inspection of reactor internals. If the catalyst is not to be unloaded and if scheduled maintenance will delay completion of the turnaround, the reactor system should be purged with nitrogen. The catalyst in the reactors should be held under 25 psig nitrogen until regeneration is to proceed.

If there is any indication that the catalyst in a reactor is not getting good flow distribution, the catalyst should be unloaded and then screened before reloading. In one instance, a large lump of agglomerated catalyst was found in the bottom of the No. 4 reactor when the catalyst was unloaded. Subsequent analysis showed 61 wt % carbon and the presence of delta alumina. Since the transition temperature from gamma to delta alumina is about 1,600°F, there is no question that the catalyst at that location had been subjected to very high temperatures.

The carbon burn of reforming catalyst does a good but incomplete job of carbon removal. Table 5–20 shows analyses of catalyst removed from three different reformers after a carbon burn. The variations within a unit are the result of bypassing catalyst or variations in sampling as the catalyst was unloaded.

TABLE 5–20
Carbon on catalyst after a carbon burn

	Carbon on Catalyst, wt %			
Reactor Number	1	2	3	4
Unit A	1.52	1.55	0.21	0.12
Unit B	0.21	0.72	3.45	1.99
Unit C	0.03	0.92	2.26	>6.60

To ensure that a catalyst is thoroughly cleaned of carbon, the second step in regeneration is the cleanup or proofburn. Unless the catalyst is to be cooled and unloaded, this step immediately follows the carbon

134 CATALYTIC REFORMING

burn. The reactor inlet temperatures are gradually raised to 900–950°F, until reactor delta T's are essentially zero. The oxygen concentration is gradually raised to 5–6 mol % in the regeneration gas. When observation of delta T's and oxygen in the outlet gas of the last reactor indicates no carbon burning, the catalyst is ready for the rejuvenation step.

REJUVENATION OF CATALYST

The development of catalyst rejuvenation has made regeneration a part of normal reforming operation. The significant contribution of rejuvenation is that it redisperses the agglomerated metal or metals of the catalyst. This metal redistribution essentially restores the catalyst to fresh-catalyst condition. Reforming catalysts are susceptible to sintering. The larger metal crystallites greater than 10 angstroms (Å) in diameter (chapter 4) do not have as good a reforming activity as those of less than 10-Å diameter. The agglomeration of metals during reforming is generally caused by sulfur, coke, and high temperatures.

Rejuvenation is accomplished by contact of the catalyst with a gaseous mixture of oxygen and chloride at an elevated temperature. The procedure is to continue circulation of the regeneration gas, containing 5–6 mol % oxygen, and to inject chloride for a period of 2–5 hr while maintaining catalyst at 900–950°F.

REDUCTION

After rejuvenation the metals on the catalyst are in the oxidized state and must be reduced to the metallic state with hydrogen to activate the catalyst for reforming. Reduction temperature is generally 800–850°F. Oxygen must be purged from the system before admitting hydrogen. The catalyst is cooled by the circulating gas. The unit is depressured and purged with nitrogen to remove all traces of oxygen. The catalyst is then reduced with hydrogen at 800–850°F.

The usual practice is to use electrolytic hydrogen for catalyst reduction, but refiners are beginning to question its necessity. Some think hydrogen from another reformer or hydrogen unit of 90–96% hydrogen purity serves the purpose.[40] The unknown quantity seems to be the effect on catalyst cycle life. The solution is to try it both ways and compare cycle lives.

SULFIDING AND START-UP

After reduction the catalyst is sulfided by addition of mercaptan, disulfide, or hydrogen sulfide to the circulating hydrogen. The purposes of sulfiding are to deactivate superactive catalyst sites and to avoid excessive hydrocracking when the hydrocarbon feed is started. When sulfiding is complete, evidenced by an increase of hydrogen sulfide in the recycle gas, the hydrocarbon feed is introduced with reactor temperatures in the range of 850–900°F. Moisture in the system is usually above 800 mol ppm in the recycle gas. Because of this moisture, chloride injection is started along with the hydrocarbon feedstock. It sometimes requires from two to four days for water in the recycle gas to decrease sufficiently so reforming severity can be raised. Reactor inlet temperatures are not recommended to be raised above 900°F to reach octane number until the moisture in the recycle gas is less than 50 mol ppm. A reformer takes about a week to line out (reach stable operation), during which time water and chloride adjustments are made.

The time required for each step in the regeneration procedure varies widely from unit to unit. Table 5–21 lists the data from the regenerations of three different commercial units, each with four reactors but with total catalyst loads varying from 40,000 to 94,000 lb. The time for each procedure approximates the actual time for each step, not including preparation time. The chemicals and utilities used are also listed.

TABLE 5–21
Regeneration of reforming catalyst

Unit	A	B	C
Time Required			
Carbon Burn, hr	48	84	64
Proof Burn, hr	7	38	12
Rejuvenation, hr	57	26	38
Reduction, hr	12	9	16
Sulfiding, hr	3	—	2
Chemicals Used/lb Catalyst			
Liquid Oxygen, scf	7	15	—
Liquid Nitrogen, scf	29	24	30
Hydrogen, scf	3	1	1
Chloride, lb	0.01	0.02	0.005
Sulfide, lb	0.002	—	0.001
Utilities Used/lb Catalyst			
Fuel, BTU	64,500	35,500	89,600
Steam, lb	41	—	—

The total turnaround time depends on maintenance done over and above catalyst regeneration. The three regenerations of Table 5–21 either unloaded and screened the catalyst after carbon burn or suspended regeneration after carbon burn, to take care of other maintenance items on the unit. Experience indicates that a unit with up to 100,000 lb total catalyst can be regenerated in about three to four days, feed out to feed in, if the turnaround is regeneration only.

136   CATALYTIC REFORMING

Since reformers now operate with hydrotreated feedstock, catalyst replacements occur only once every ten years or so. Because of this and more frequent regenerations, refiners are faced with the decision of how long to leave a reformer in service before unloading catalyst and inspecting reactor internals. The response to a question indicated that there is no hard and fast rule, but every two years or every third or fourth regeneration was a consensus.[41] In fact, experience shows that many times some problem arises within a two-year period that requires opening the reactors. Also, correcting poor flow distribution and cleaning plugged screens or scallops are significant benefits for the two-year interval.

MERCHANT REGENERATION

Merchant regeneration of catalyst means shipping the catalyst outside the refinery for regeneration. In the case of reforming catalyst, merchant regeneration is a misnomer. As far as is known, those in that business can only perform a carbon burn and screening. None is equipped to carry out the rejuvenation procedure using oxygen and chloride at elevated temperatures.

However, it is sometimes convenient for a refiner to have a reforming catalyst processed through a carbon burn and screening. In one instance, a bimetallic catalyst was to be replaced with a newer, more active one. The bimetallic catalyst was still capable of years of good service. It was merchant regenerated, screened, and used for catalyst makeup in other units.

Although merchant regeneration requires more investment in precious metals, having a spare load of catalyst already carbon burned and screened shortens turnaround time. In the author's experience, merchant carbon burn of bimetallic reforming catalyst is entirely satisfactory. During one such burn, samples of catalyst were taken at the entrance, the midpoint and the discharge of the regeneration unit, with the following results:

	Feed	Midpoint	Discharge
Carbon, wt %	15.30	6.10	0.30
Chloride, wt %	1.01	0.77	0.50

A maximum temperature of 900°F during the burn was specified. There was concern that some platinum and rhenium might be lost. Many samples were assayed, and both the catalyst supplier and the refiner were satisfied that there was no evidence of loss of either metal. The catalyst fines were recovered from screening, and the refiner noted that there was no significant loss of platinum or rhenium from the time the catalyst left the plant until its return.

PROCESS VARIABLES AND UNIT OPERATION

Merchant carbon burning is almost always more thorough than that done in situ in the reformer reactors. The catalyst is screened before and after the merchant regeneration unit. The units may be a specially designed rotary kind or a moving-belt type. Either does very good jobs of temperature control and of carbon removal.

SAFETY

Catalytic reformers are designed for maximum safety of operating personnel and for protection of catalyst and equipment. One aspect of reforming that generates concern for safety is the presence of large volumes of hydrogen at high pressure. The explosiveness of hydrogen was manifested by the Hindenberg dirigible fire.

Hydrogen

Hydrogen is a flammable gas. It is colorless, odorless, tasteless, and nontoxic. At atmospheric pressure the ignition temperature of hydrogen-air mixtures is reported to be as low as 932°F (500°C).[42] The most serious hazard of hydrogen is its wide range of flammable limits in air, of 4.1–74 vol % (Fig. 5–21). By comparison, flammability limits

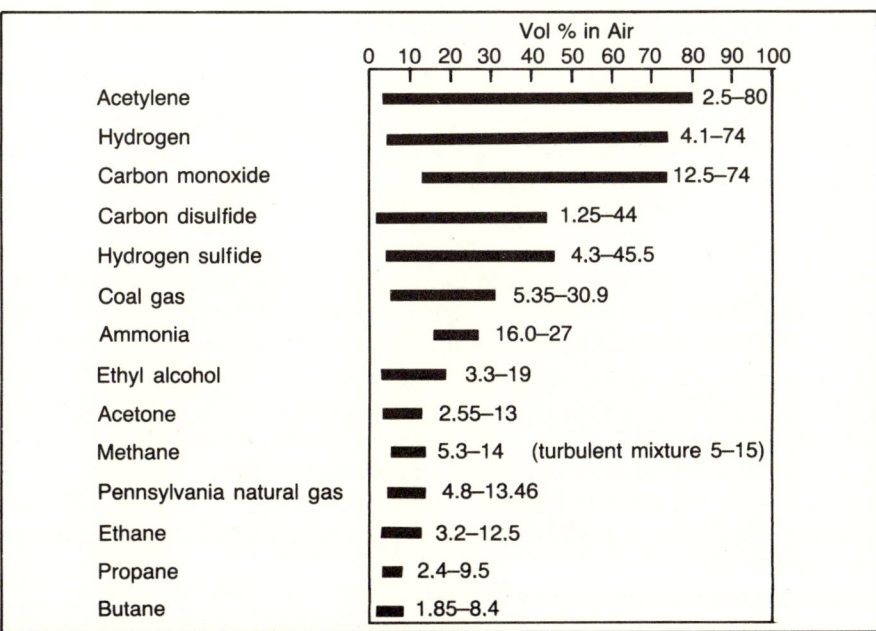

Fig. 5–21 Limits of flammability of single gases and vapors (adapted from National Fire Protection Assoc. data and other sources)

of natural gas are 4.8 and 13.46; ethane 3.2 and 12.5; propane, 2.4 and 9.5; and butane, 1.85 and 8.4 vol % of gas in air.

Hydrogen burns in air with a pale blue flame which is nearly invisible in daylight. Leaks of hydrogen-containing streams on a reformer unit are often leaking flanges of exchangers or piping. Operators are aware of this hazard and watch for hydrogen fires, especially during startups.[43]

A far more serious occurrence is the failure or rupture of equipment or lines, releasing large volumes of hydrogen into the atmosphere. Operators are trained to handle such emergencies. The usual procedure is to shut off the fires in the heaters, shut down the reactor charge pump, and depressure the unit to the flare.

Other emergencies, which do not release hydrogen to the atmosphere, are loss of instrument air, electricity, or steam. When these occur, circulation of hydrogen to protect the catalyst is continued if possible.

Temperature Runaway

One safety hazard which has been reported more than once is overheating of the reactor shell, with consequent bulging of the shell or failure to the point of splitting.[44] Two possible causes of extreme reactor-shell temperature have already been mentioned: a) too-high percentage of oxygen in regeneration gas and b) demethylation reaction near the reactor shell.

Radial-type reactors are particularly susceptible to temperature excursions, a result of catalyst accumulation in the annulus, between the shell and the catalyst-retaining screen. In such circumstances, vapor flow and space velocity are so low that exothermic reactions of hydrocracking and demethylation produce very high temperatures. The area is stagnant, so there is little dispersion of heat to flowing gas. As a safety device, many radial flow reactors are equipped with several thermocouples peened* at intervals into the reactor shell bottom and up to about three feet above the bottom tangent line. The thermocouples are continually scanned. An alarm sounds if the temperature at any point rises above a certain maximum.

An example of an actual temperature excursion in the No. 2 reactor of a commercial reformer is shown in Fig. 5–22. The reactor-skin couple recorded a temperature rise of 350°F in 12 hours. Shutdown of the unit began when the skin thermocouple recorded a shell temperature of 1,150°F. When the reactor was inspected, a hole was found in the catalyst retaining screen. Twelve feet of catalyst and iron sulfide scale were found in the annulus. The catalyst in the bottom of the reactor was severely coked and fused into a solid mass.

*Method for attaching thermocouple wire to a reactor

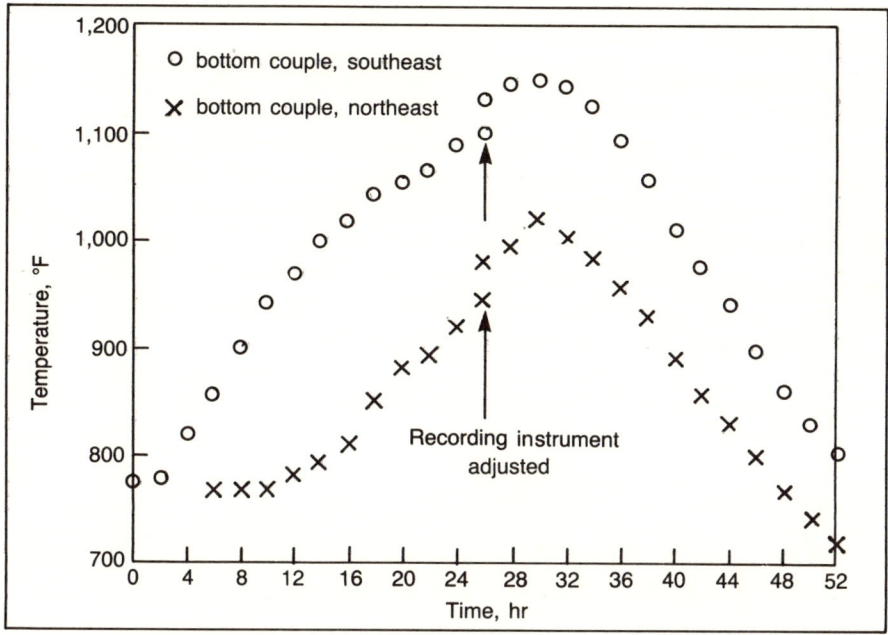

Fig. 5–22 Number 2 reactor temperature excursion

This reactor was of the hot-wall type with external insulation. Cold-wall reactors with internal insulation also report reactor shell hot spots. The cold-wall reactor shells are monitored by painting the outside of the shell with temperature-sensitive paint. Downflow reactors are usually not equipped with thermocouples on the reactor shell.

SUMMARY

The performance of a catalytic reformer depends upon a combination of controllable variables. Reactor temperatures, space velocity, chloride/water balance, and level of feed contaminants all influence catalyst activity, selectivity, and stability. When the catalyst is working properly, the reformer is performing at its best, that is, putting out maximum yield of the desired product.

Constant surveillance of reformer operation is the key to maintaining peak performance. Daily review and plotting reactor delta T's, hydrogen purity in separator gas, butane-and-lighter yield, and the reformate yield and its octane number give indications of unit performance. These data can also alert a technologist to changes that may signal trouble. A weekly or biweekly material weight balance including calculated reformate yield deviation from that of fresh catalyst is essential. Recognition of signs of catalyst poisoning by sulfur, nitrogen, lead, and arsenic, followed by quick corrective measures, can prevent an expensive catalyst regeneration or replacement.

Reformers are subject to fouling and corrosion in heaters, exchangers, piping, and vessels. Recognition of cause and control of these problems is essential to avoid equipment failures. Safety is most important in the operation of a catalytic reformer, especially an awareness of the presence of high-pressure hydrogen.

REFERENCES

1. NPRA 1983 Q&A on Refining, p. 74, question 43.
2. Aalund, Leo R. "Guide to Export Crudes for the 80s." Reprinted from a series of articles in *OGJ*, 11 April–19 December 1983. © 1983 by PennWell Publishing Co.
3. NPRA 1969 Q&A, p. 65, question 27.
4. NPRA 1977 Q&A on Refining, p. 77, question 3.
5. NPRA 1978 Q&A on Refining, p. 92, question 46.
6. NPRA 1978 Q&A on Refining, p. 91, question 44.
7. 1982 Ketjen Catalysts Symposium, Reforming Operations, p. 31, question F–3.
8. NPRA 1976 Q&A on Refining, p. 24, question 2.
9. NPRA 1981 Q&A on Refining, p. 99, question 6.
10. D'Auria, James, William C. Tieman, and George Antos. "Recent Platforming Catalyst Developments." Paper AM–80–49 presented at the annual meeting of NPRA, 1980.
11. Figoli, N.S., et al. "Operational Conditions and Coke Formation on Pt/Al_2O_3 Reforming Catalyst." *Applied Catalyses* 5 (14 January 1983): 19–32.
12. Edgar, M. Dean. "Catalytic Reforming of Naphtha in Petroleum Refineries." In *Applied Industrial Catalysis* 1. Academic Press Inc. 1983.
13. NPRA 1975 Q&A, p. 26, question 9.
14. NPRA 1978 Q&A, p. 92, question 48; pp. 93–94, question 50.
15. NPRA 1977 Q&A, p. 93, question 28.
16. Eng, Jackson, and Janis Bumbulis. U.S. pat. 3,434,961.
17. NPRA 1975 Q&A, p. 31; NPRA 1976 Q&A, p. 26; NPRA 1977 Q&A, p. 90; NPRA 1978 Q&A, p. 94; NPRA 1982 Q&A, p. 100, op. cit.
18. NPRA 1975 Q&A, p. 29, question 14.
19. NPRA 1975 Q&A, p. 30, question 15 (added comment).
20. NPRA 1981 Q&A, p. 96, question 21.

21. NPRA 1975 Q&A, p. 126.
22. Technical memorandum no. 303, DuPont Co.
23. NPRA 1978 Q&A, p. 87, question 35.
24. NPRA 1982 Q&A, p. 99, question 7.
24. NPRA 1974 Q&A, p. 24; NPRA 1976 Q&A, p. 27; NPRA 1977 Q&A, pp. 81, 87; NPRA 1978 Q&A, pp. 84, 95; NPRA 1979 Q&A, pp. 84, 86; NPRA 1981 Q&A, p. 92.
26. NPRA 1975 Q&A, p. 28; NPRA 1976 Q&A, p. 16; NPRA 1978 Q&A, p. 75; NPRA 1979 Q&A, p. 86; NPRA 1981 Q&A, p. 94; NPRA 1982 Q&A, p. 97.
27. May 1982 Ketjen Catalysts Symposium, p. 32.
28. Gladrow, Elroy M., et al. "Catalytic Cracking Process." U.S. pat. 3,793,192.
29. Huling, G.P., and J.D. McKinney. "Feed Sulfur Distribution in FCC Product." *OGJ* (19 May 1975): 73–79.
30. Kress, R.F., H.W. Nagel, and H.E. Reif. "Reforming Heavy Catalytically Cracked Gasoline for High Octane Blend Stock." Chemical Engineering Progress Symposium Series 57 (34): 8–13. Published in *Petrochemicals and Petroleum Refining*, 1961.
31. NPRA 1976 Q&A, p. 30, question 14.
32. Bailor, et al. "Catalyst Regeneration Process." U.S. pat. 4,406,775. 27 September 1983.
33. Kearby, et al. "Reactivation of Regenerated Noble Metal Catalysts with Gaseous Halogens." U.S. pat. 3,134,732. 26 May 1964.
34. NPRA 1977 Q&A, p. 91, question 24.
35. NPRA 1978 Q&A, p. 79, question 18; NPRA 1976 Q&A, p. 21, question 8; NPRA 1981 Q&A, p. 98, question 25.
36. NPRA 1977 Q&A, p. 86, question 13.
37. NPRA1978 Q&A, pp. 77–78, question 17.
38. Ibid.
39. NPRA 1979 Q&A, p. 81, question 32.
40. NPRA 1978 Q&A, p. 88, question 38.
41. NPRA 1978 Q&A, p. 91, question 45.
42. "Standard for Gaseous Hydrogen Systems at Consumer Sites." 1978 Edition of the National Fire Protection Assoc., NFPA 50A–1978.
43. NPRA 1982 Q&A, p. 102, question 34.
44. NPRA 1981 Q&A, p. 97, question 24; NPRA 1982 Q&A, p. 102, question 35.

CHAPTER 6

BTX Operation

Benzene, toluene, orthoxylene, metaxylene, and paraxylene are valuable chemicals. The derivatives from these basic materials make thousands of products found in every home and office. The major endproducts of these chemicals are listed below.

- **Benzene**—source of styrenic plastics, phenolic polymers, and nylon
- **Toluene**—a major source of benzene through a hydrodealkylation process; also useful as a solvent in paints and adhesives
- **Orthoxylene and paraxylene**—sources of polyester fibers, plastic film, and plastic bottles
- **Metaxylene**—usually converted to paraxylene through catalytic isomerization processes.

The major source of benzene, toluene, and xylenes (BTX) is catalytic reforming; it supplies over 80% of these key chemicals for the United States. A second and increasing source of BTX is pyrolysis naphtha from naphtha cracking. A minor BTX source is gas from coke ovens.

Catalytic reforming for *aromatics* (benzene, toluene, and xylenes) differs little from reforming to raise octane number for motor fuel blending. The highest octane number reformates have the highest concentration of aromatics.

One study showed C_5^+ reformate of 100 RON clear contained about 72 vol % aromatics.[1] This does not mean the yield of aromatics from feed is 72 vol %. The aromatics yield depends on many variables (see Equilibrium Distribution of Aromatics). A C_5^+ reformate of 100 RON clear has an aromatic content (C_6, C_7, C_8, and C_9^+) of about 72 vol %.

The same study reported a C_5^+ reformate of 107.5 RON clear (the highest reformate octane number I have seen) contained 88.1 vol % aromatics. A BTX reformer and a motor fuel reformer are operated essentially for the same reason, i.e., to yield aromatics.

FEEDSTOCK FOR BTX REFORMING

It is advantageous for a refiner producing BTX chemicals to have two reformers, one for BTX production and one for increasing motor-fuel octane number. A single reformer will certainly yield benzene, toluene, and xylenes; the reason for two reformers is selectivity. Refiners recognize that proper selection of feedstock maximizes yield of the desired product.

For example, cat cracking gas oil and topped crude in separate units produces more gasoline than cracking them combined in one unit. Also, desulfurizing naphtha and distillate in separate units is better than desulfurizing them in one HDS unit. In this case, the distillate is treated at a higher severity than necessary because the naphtha must be desulfurized to meet reformer feed specification. A separate reformer for BTX production is best because

- Feed stock naphtha can be selected for maximizing yields
- A BTX reformer charging lower-boiling hydrocarbons can operate at a lower pressure and a higher temperature without excessive catalyst coking than can the higher-boiling hydrocarbons charged to a motor-fuel reformer
- The heavy components in motor-fuel naphtha show less selectivity to aromatics at low pressure; aromatics from heavy components tend to be C_9 and heavier (C_9^+) instead of BTX.

EQUILIBRIUM DISTRIBUTION OF AROMATICS

The chemical reactions of chapter 2 were studied extensively to determine the best operating conditions for maximizing aromatics in BTX reforming. Two approaches are made to understanding the effects of temperature, pressure, and hydrogen partial pressure on aromatics yield. One is the calculated thermodynamic equilibrium of reactants and products.[2,3,4,5,6] An example of calculated equilibrium for specific hydrocarbons is shown in Fig. 6–1 for the n-heptane–toluene–hydrogen system.

The other is pilot unit tests on individual hydrocarbons or selected naphthas.[7,8,9,10,11,12,13,14,15] Two examples of equilibrium by pilot unit tests are illustrated in Figs. 6–2 and 6–3. For the tests of Fig. 6–2, feed naphthas of 100%, 66%, and 39% paraffins were reformed; total aromatics in the C_5^+ reformate are plotted against RON clear. In Fig. 6–3, the composition of C_5^+ reformate at increasing RON clear is plotted as paraffins and naphthenes. Aromatics are obtained by difference: at 100 RON clear, aromatics content is about 71.2 vol %. In the tests of Fig. 6–2, the percent aromatics in 100 RON0 C_5^+ reformate from 100% paraffin feed approaches that from the 66%- and the 39%-paraffin feeds.

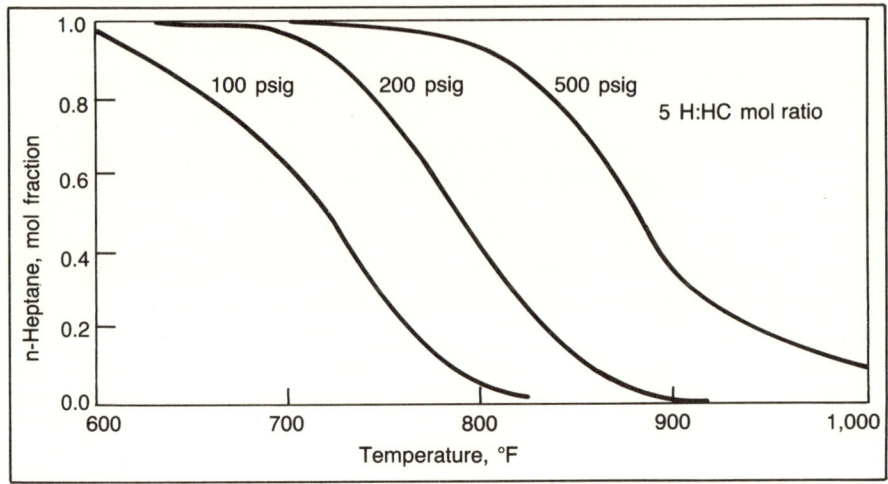

Fig. 6–1 Equilibrium distribution for system: n-heptane–toluene–hydrogen *(after Hettinger, Keith, Gring, and Teter, ref. 3; courtesy American Chemical Soc.)*

The aromatics from 100%-paraffin feed could only have come from dehydrocyclization of paraffins. However, high-octane components include lower-boiling paraffins from hydrocracking and isomerization. Also, elimination of low-octane paraffins from C_5^+ reformate by hydrocracking them to butanes and lighter hydrocarbons contributes to higher octane reformate.

It would be convenient to have correlations for estimating yields of individual aromatics such as benzene, toluene, and xylenes, based on feedstock composition—something similar to the N + 2A correlations for motor-fuel reforming. Except for a few theoretical models based on thermodynamics, existent reforming models are proprietary to licensors, catalyst manufacturers, and petroleum refiners who developed them.[17] A few engineering companies market reforming computer models (software). For application to an existing unit, such models are usually modified by "unit factors" derived from operating data and test runs. The following generalizations were derived from published literature, pilot unit tests, and commercial reforming experience:

- Maximum aromatics yield is favored by low pressure and high temperature.
- Dehydrocyclization of paraffins is more favored by lowering pressure than by raising temperature. Higher temperatures promote hydrocracking. Modern reformers, designed for 50–150-psig re-

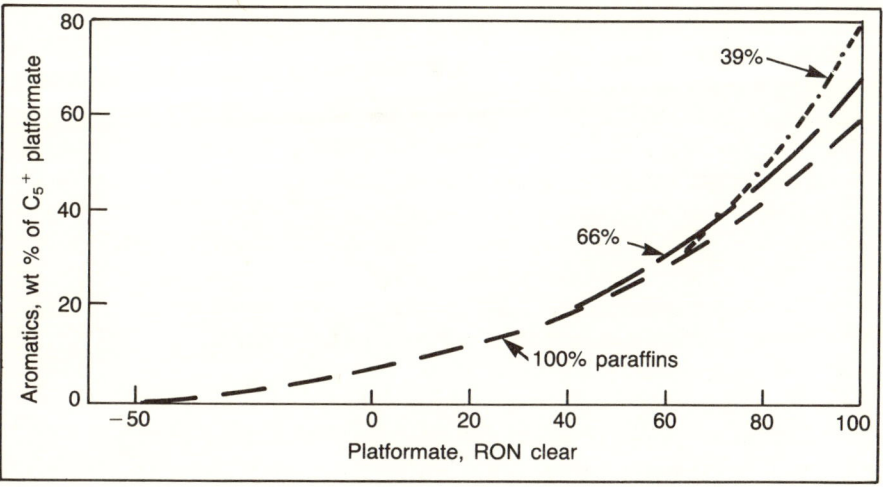

Fig. 6–2 Aromatics, weight percent of C_5^+ platformate *(after Donaldson, Pasik, and Haensel, ref. 16; courtesy American Chemical Soc.)*

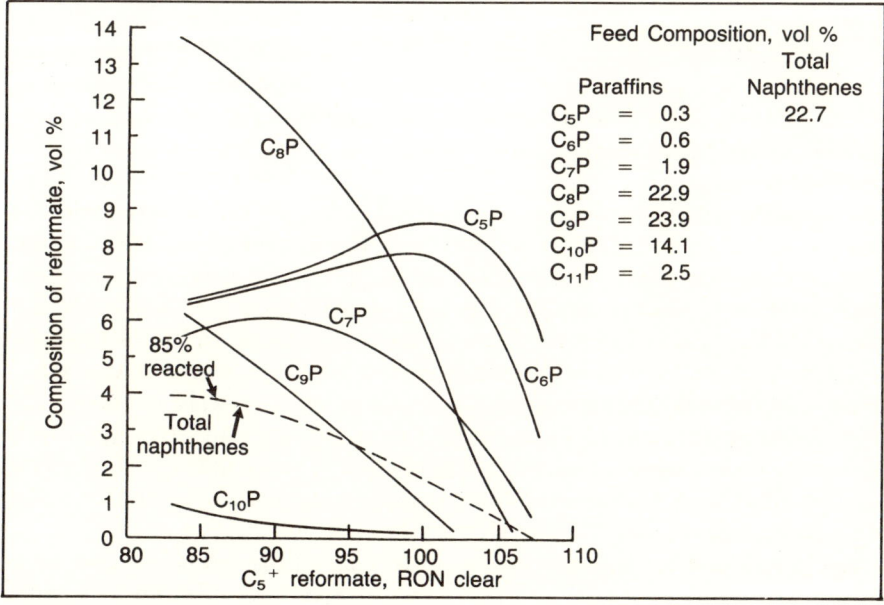

Fig. 6–3 Effect of severity on composition of reformate, heavy Arabian feed *(after Jacobsen, Hughes, and Robinson, ref. 1; courtesy Chevron Research Co.)*

146 CATALYTIC REFORMING

actor pressure, can convert a high percentage of paraffins to aromatics.
- Hydrogen partial pressure affects equilibrium as well as total pressure—one reason for decreasing H:HC ratios.
- The boiling range of the feedstock affects the type of aromatic produced. Benzene precursors are in the 160–180°F boiling range. Toluene precursors are in the range of 180–250°F. Xylene precursors boil at ranges from 250 to 350°F. Hydrocarbons boiling above 350°F generally yield C_9^+ aromatics.
- Cracked naphthas are a rich source of aromatics and naphthenes. After hydrotreatment they are premium feedstocks for BTX units (see chapter 5.)[18]
- The higher the aromatic content of the feed to a reformer, the higher the yield of aromatics based on feed. However, if a feed is rich in aromatics, more total barrels of aromatics can be produced by extracting the aromatics from the feed and reforming only the naphthenes and paraffins.
- Aromatics may not always be the most profitable product to produce. For example, toluene may have a higher value as a motor-fuel component than as a petrochemical.

PROCESS VARIABLES

Process variables for BTX reforming are the same as for motor-fuel reforming. Reactor inlet temperature, reactor delta T, reactor pressure, H:HC ratio,[19] and space velocity are monitored as they are for motor-fuel reforming.

Selected data from a commercial BTX reformer material balance are shown in Table 6–1. The reactor-charge boiling range was 184–314°F. Average reactor pressure was 250 psig. The C_5^+ reformate octane number (unleaded) was 97.0 for the day this material balance was obtained. Note that chromatographic analyses were reported for paraffins, for naphthenes, and for aromatics through C_9 hydrocarbons. From these data, together with reactor temperature and pressure, a computer model calculated the predicted yields of benzene, toluene, and C_8^+ aromatics for comparison with actual yields (Table 6–1). The predicted and actual yields were close for benzene and toluene but were off by more than 1% for C_8^+ aromatics.

A series of pilot unit tests on a highly paraffinic (89.5%) raffinate, a heavy straight-run naphtha (54.6% paraffins), and a combined 20%-raffinate–plus–80%-naphtha feedstock is illustrated in Fig. 6–4. The results show conversion of C_7 paraffins to toluene by dehydrocyclization and the hydrocracking of C_7 paraffins to C_4 and lighter hydrocarbons,

TABLE 6-1
Commercial BTX reformer material balance data

	Reactor Charge, liquid vol %	Reformate, liquid vol %
Isopentane	0.14	0.29
Normal pentane	0.23	0.55
C_6 paraffins	5.11	9.11
Methylcyclopentane	1.92	0.48
Cyclohexane	2.53	0.04
Benzene	1.19	5.00
C_7 paraffins	16.72	14.35
C_7 naphthenes	12.81	0.55
C_7 aromatics	4.66	21.14
C_8 paraffins	17.70	6.63
C_8 naphthenes	12.33	0.95
C_8 aromatics	6.70	27.42
C_9 paraffins	9.62	1.20
C_9 naphthenes	7.66	0.17
C_9 aromatics	0.68	12.12
Total	100.00	100.00
Paraffins	49.5	
Naphthenes	37.2	
Aromatics	13.3	
N + 2A	63.8	

Aromatics Correlation	Liquid Vol % in Feed	Predicted Aromatic Yield, liquid vol % Feed	Actual Aromatic Yield, liquid vol % Feed
Benzene Precursors			
C_6 paraffins	5.11	0.26	
Cyclohexane	2.53	2.01	
Methylcyclopentane	1.92	0.69	
Benzene	1.19	1.19	
Total benzene		4.15	4.10
Toluene Precursors			
C_7 paraffins	16.72	3.23	
Methylcyclohexane	7.91	6.51	
Dimethylcyclopentane	4.90	1.79	
Toluene	4.66	4.66	
Total toluene		16.19	17.06
C_8^+ Aromatic Precursors			
C_8^+ paraffins	27.31	9.10	
C_8^+ naphthenes	19.98	16.71	
C_8^+ aromatics	7.38	7.38	
Total C_8^+ aromatics		33.19	31.90

148 CATALYTIC REFORMING

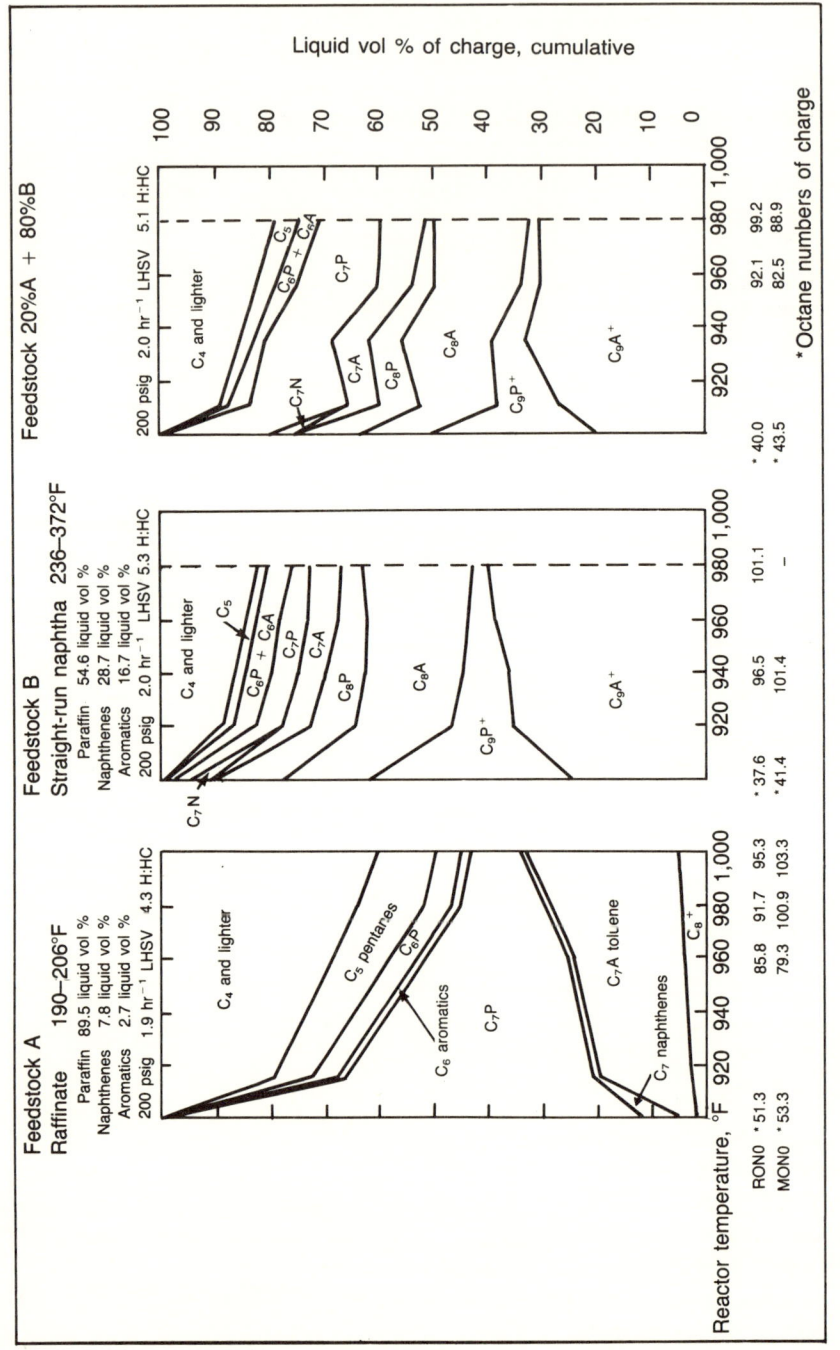

Fig. 6–4 Pilot unit reforming of raffinate and straight-run naphtha

as reactor temperatures approached 1,000°F. A study of the data indicated a slight advantage in C_5^+ reformate yield when reforming the raffinate and naphthas individually, that is, not combined.

Another example of the effect of feedstock composition on aromatics yield is Fig. 6–5. The highest yield of aromatics is from Isomaxate feed which contained only 32.6 vol % paraffins. The lowest yields of aromatics are from light Arabian naphtha feed which contained 74.6 vol % aromatics. These tests indicated the aromatics yield based on feed is near a maximum when the C_7^+ paraffins have almost all reacted. The small arrows on Fig. 6–5 indicate the octane level at which 6% and 3% C_7^+ paraffins remain in the reformate.

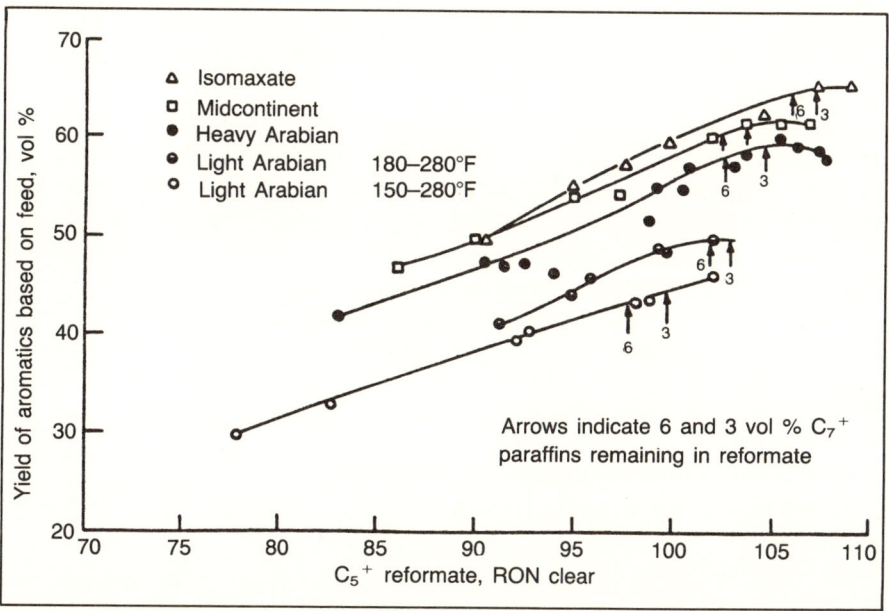

Fig. 6–5 Aromatics vs severity, 200-psig rheniforming *(after Jacobsen, Hughes, and Robinson, ref. 1; courtesy Chevron Research Co.)*

Although computer models are available and equilibrium calculations are made, nearly all refiners run a daily octane number on C_5^+ reformate from BTX operation. In fact, refiners often operate the BTX reformer with the objective of producing a certain-octane-number reformate. Others follow the conversion of a certain hydrocarbon as the objective. For example, a BTX reformer may be operated to convert 75–80% of the methylcyclopentane in the feed.

The bottom line is that much progress has been made, but refiners still have a long way to go in commercial BTX reforming.

LABORATORY ANALYSIS FOR AROMATICS

It is one thing to lower reactor pressure and to raise temperature to yield more aromatics; it is another to accurately determine the aromatics content of the reformate. The market value of aromatic hydrocarbons is normally such that a few tenths of a percent variance can result in thousands of dollars' difference in refinery income.

During a series of pilot unit tests to compare aromatic yields from different catalysts and to determine the effects of process variables, establishing the accuracy of laboratory analyses for aromatics became imperative. A scheme was devised for preparing a reformate of known aromatics composition and for submitting this reformate for analysis to a number of different laboratories.

The standard reformate was prepared from a BTX reformate by extracting aromatics from this BTX reformate with sulfuric acid until the raffinate was completely free of aromatics, as evidenced by nuclear-magnetic-resonance (n-m-r) and gas-chromatograph (gc) analyses. A separately prepared quantitatively known blend of aromatics was added to the raffinate. The weights of each were known, so the concentration of aromatics in the blend was known. The synthetically prepared sample was submitted to six laboratories (representing licensors, catalyst suppliers, and plant operations).

The results of the analyses are in Table 6–2. The laboratories were not asked to reveal analytical methods used, but gas chromatography was probably the predominant method. Judging the merits of the laboratories is not the point of giving these data. These numbers are presented to show the range of accuracy of analyses for aromatics in reformate in the early 1980s.

SUMMARY

Catalytic reforming for aromatics (BTX) follows the same principles as reforming for motor fuel (raise octane number). Feedstocks are selected to maximize yields of benzene, toluene, and xylenes. The aromatics must be removed from the reformate, usually by extraction, and purified to specification for marketing.

Process variables and operating parameters for BTX reforming have been developed by refiners and licensors of BTX technology. This information is proprietary. Except for a few published articles on pilot unit testing or on thermodynamic equilibrium of reactants and products, little information is available.

TABLE 6–2
Comparative analysis of synthetic reformate (wt %)

Components	Synthetic Reformate	Laboratory					
		A	B	C	D	E	F
Benzene	5.41	5.20	6.3	5.06	5.74	5.14	6.1
Toluene	21.40	21.43	23.4	21.41	22.52	21.63	23.3
Ethyl benzene	4.28	4.52	4.2	4.32	4.03	4.43	4.4
p-Xylene	4.26	^{a}{15.45	4.2	4.24	3.95	^{a}{15.73	3.9
m-Xylene	10.69		10.6	10.84	9.89		11.0
o-Xylene	5.44	5.82	6.4	5.51	5.00	5.85	5.5
Isopropyl benzene	1.07	1.25	1.1	1.00	1.05	1.22	—
n-Propyl benzene	1.07	1.15	1.1	0.95	1.03	1.11	—
1-Me, 3-ET benzene	1.06	^{a}{2.19	^{a}{2.1	^{a}{2.01	1.07	2.16	—
1-Me, 4-ET benzene	1.04				0.95	—	—
1-Me, 2-ET benzene	1.08	1.15	1.1	1.02	1.04	1.27	—
1,3,5-Tri-me-benzene	1.07	1.15	1.1	0.99	1.06	1.14	—
1,2,4-Tri-me-benzene	2.16	2.33	2.2	2.08	2.06	2.53	—
1,2,3-Tri-me-benzene	1.11	1.22	1.0	1.01	1.02	1.19	—
Total C_8 aromatics	24.67	25.79	25.4	25.02	22.87	26.01	24.8
Total C_9 aromatics	9.66	10.44	9.7	9.06	9.28	10.62	7.9
Total aromatics	61.13	62.86	64.8	60.55	59.41	63.40	62.1

^{a}Combined value of unresolved pair

REFERENCES

1. Jacobsen, R.L., T.R. Hughes, and R.C. Robinson. "Economics of High Severity Rheniforming and of Reformer Feed Initial Boiling Point Selection." Paper presented at the 38th meeting of the API's Div. of Refining in Philadelphia, 15 May 1973.
2. Ciapetta, Frank G. "Catalytic Reforming." *Petro/Chem Engineer* (May 1961): C19–C31.
3. Hettinger Jr., W.P., C.D. Keith, J.L. Gring, and J.W. Teter. "Hydroforming Reactions." *I&EC* 47 (April 1955): 719–730.
4. Kirkbride, C.G. "Houdriforming—A Continuous Process for Reforming Petroleum Naphthas." *Petroleum Refiner* (June 1951).
5. Kugelman, Alan M. "What Affects Cat Reformer Yield?" *Hydrocarbon Processing* (January 1976): 95–102.
6. Kopf, F.W., W.H. Decker, W.C. Pfefferle, N.H. Dalson, and J.A. Nevison. "Magnaforming Offers Greater Yields." *OGJ* (19 May 1969): 141–152.
7. Jacobsen, Hughes, and Robinson, op. cit.
8. Kopf, Decker, Pfefferle, Dalson, and Nevison, op. cit.
9. Hettinger, Keith, Gring, and Teter, op. cit.
10. Pollitzer, E.L., J.C. Hayes, and V. Haensel. "The Chemistry of Aromatics Production via Catalytic Reforming." In ACS Symposium on Refining Petroleum for Chemicals, D8–D17. New York City: 7–12 September 1969.
11. Addison, G.E., J.E. Conway, and P. Kuchar. "Aromatics Opportunities Are Plentiful for U.S. Refiners." *OGJ* (7 July 1975): 56–66.
12. Skipin, Yu A., A.A. Yakolev, and G.A. Gorodetskaya. "Increasing the Output of Aromatic Hydrocarbons in Catalytic Reformers." © Plenum Publishing Corp. *Khimiya: Tekhnologiya Topliv: Masel* 10 (October 1982): 10–11.
13. Fedorov, E.P., E.A. Shkuratova, L.V. Bulygina, and A.A. Potapov. "Influence of Reforming Conditions and Feedstock Distillation on Aromatization of Paraffinic Hydrocarbons." © Plenum Publishing Corp. *Khimiya: Tekhnologiya Topliv: Masel* 3 (March 1982): 16–19.
14. Haensel, Vladimir. "Aromatics Production by the Platforming Process." *OGJ* (9 August 1951): 80–84.
15. Haensel, Vladimir, and Charles V. Berger. "Aromatics by Platforming." *Petroleum Processing* (March 1951): 264–267.
16. Donaldson, G.R., L.F. Pasik, and Vladimir Haensel. "Dehydrocyclization in Platforming." *I&EC* 47 (April 1955): 731–739.
17. Kugelman, op. cit.
18. NPRA 1977, Q&A, p. 83, question 6; NPRA 1979 Q&A, p. 90, question 53.
19. NPRA 1978 Q&A, p. 92, question 48.

CHAPTER 7

Commercial Processes

In the modern refineries, catalytic reformers come in all shapes and sizes. This was not always the case. The first reformers, in the 1940s and early 1950s, were high-pressure units. To avoid rapid deactivation of the catalyst by coking, reactor pressures ranged from 400 to 700 psig. Hydrogen-to-hydrocarbon ratio ranged from 6 to 10 mols of hydrogen per mol of hydrocarbon. In comparison, reformers in 1984 operate with reactor pressures from 100 to 150 psig, at H:HC ratios of 3:1–5:1 and get much longer catalyst life than the 1940 reformers.

In the early units, if a refiner tried to increase reformate yield by lowering pressure or tried to increase reformate octane number by raising reactor temperature, the catalyst would rapidly lose activity and selectivity, due to carbon on the catalyst. It was evident to those experienced in reforming that yields could be improved and higher severity could be attained. One way was by better catalysts. Another way was a good regeneration technique. Those companies which had research and development facilities and whose management could be persuaded of the potential for a good return went to work.

The regeneration technology came first. Catalyst developments were slow, and it was the late 1960s—when the bimetallic and multimetallic catalysts appeared—that the real breakthrough occurred. As a result of intensive laboratory and pilot unit studies, a number of companies designed and built their own reformers, which also utilized proprietary catalysts. Some were offered to outsiders through licensing; others were not.

A brief survey of construction listings[1] and refining reviews[2] shows six catalytic reforming processes that are in commercial use and are available for license. The processes' licensors have courteously supplied information. Each process is summarized in this chapter in an equitable manner. The processes are listed in Table 7–1 in alphabetical order and are presented in that sequence.

TABLE 7-1
Commercial catalytic reforming processes

Process Name	Licensor	Process Type	Catalyst[a]
IFP Catalytic Reforming	Institut Francais du Petrole	Semiregenerative, moving bed	Proprietary
Magnaforming	Engelhard Industries Div. of Engelhard Corp.	Semiregenerative, semicyclic	Proprietary
Platforming	UOP Process Div. of UOP Inc.	Semiregenerative, continuous catalyst regeneration	Proprietary
Powerforming	Exxon Research and Engineering Co.	Cyclic, semiregenerative, semicyclic	Proprietary
Rheniforming	Chevron Research Co.	Semiregenerative	Proprietary
Ultraforming	Standard Oil Co. (Indiana)	Cyclic, semiregenerative, semicyclic	Proprietary
	Amoco Research and Development Dept.		

[a] More than one type of catalyst is available from each licensor. Most are bimetallic or multimetallic (see Table 4-2).

COMMERCIAL PROCESSES

Before proceeding, two important observations should be made. First, the typical utility requirements at the end of each process cannot be used for comparison of processes. Feed composition, severity, and feed rate are not the same in each case, so there is no basis for comparison. The "typical" utilities presented give an indication of the energy requirements in catalytic reforming.

Second, control of sulfur in reformer feed is important in all current catalytic reforming processes. Some processes offer a sulfur guard reactor between the HDS unit and the reformer. Others rely on the HDS unit to provide satisfactory sulfur removal. In either case, modern bimetallic or multimetallic catalysts require that the sulfur levels in the feed be very low—from below the detection limit to 1 ppm wt, depending on the catalyst used.

IFP CATALYTIC REFORMING

The *IFP* catalytic reforming process is licensed by Institut Francais du Petrole, which first developed a *semiregenerative* technology, followed a few years after by a *continuous regenerative* one. This last process (Fig. 7–1) is characterized by continuous circulation of the catalyst by the moving-bed system. Catalyst regeneration is carried out in a fixed-bed regenerator using the same procedures as the one duly proven in the semiregenerative process.

In the semiregenerative process (Fig. 7–2) a mean reactor pressure of 200 psig provides acceptable cycle length with most feedstock. With regenerative process this pressure can be reduced to as low as 110 psig to take advantage of better yields. Significant features of the IFP regenerative technology are:

1. The entire sequence of operation, including withdrawal and circulation of catalyst, is programmed on a real-time computer.
2. Catalyst is transferred from one reactor to another by means of a gas-lift system. Gas-tight valves never open or close on the catalyst. Catalyst-tight valves open but never close on the catalyst. Few valve actions are required (two or three times per day).
3. The regeneration sequence is automatic. Extensive self-checking is carried out by the computer before each step in the sequence. The entire regeneration operation is displayed visually on a console. Manual override provisions exist.
4. A small fraction of the total catalyst inventory is regenerated at a time.
5. Catalyst sampling is possible for optimum follow-up of the different operations.
6. Safety was carefully considered in designing the unit. The regen-

156 CATALYTIC REFORMING

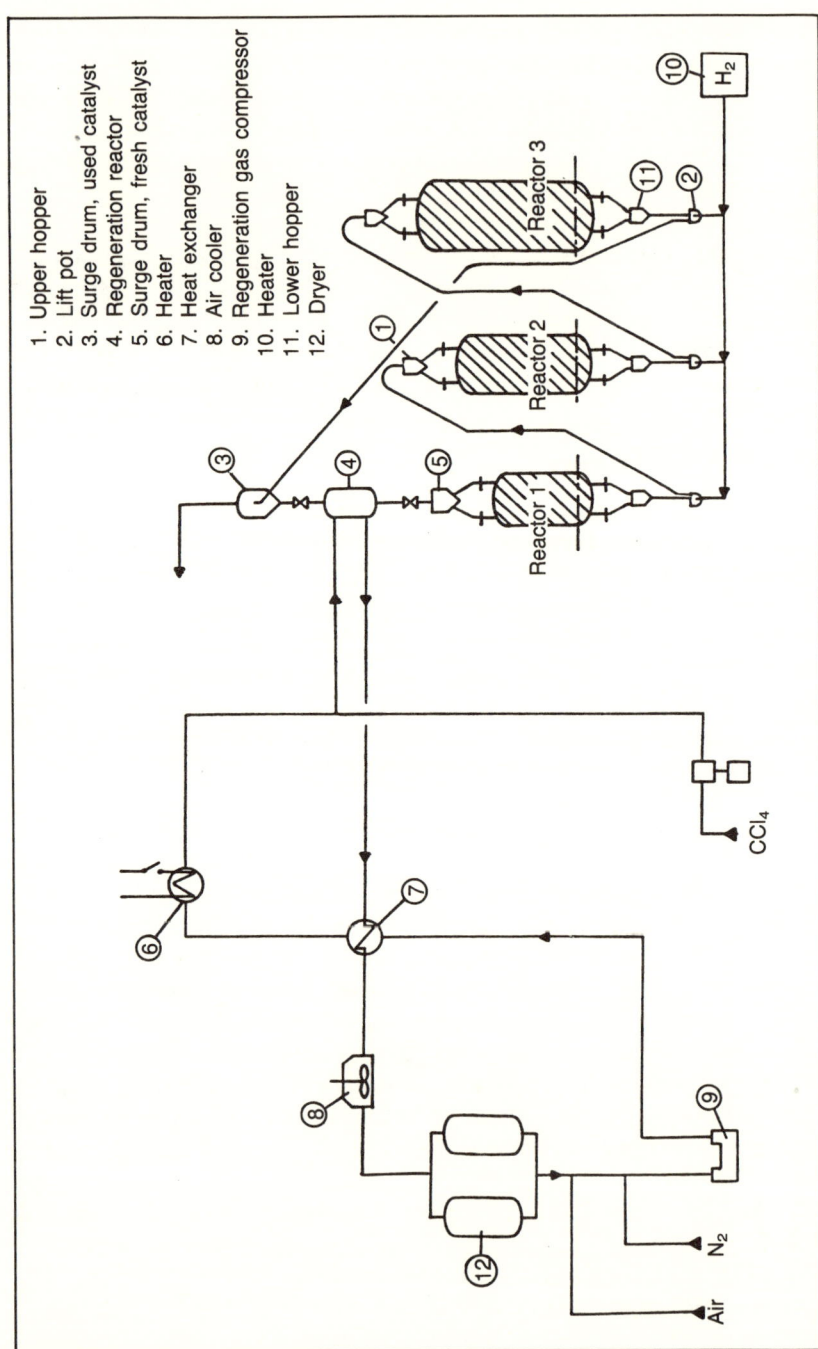

Fig. 7–1 IFP continuous regenerative system *(courtesy IFP Enterprises Inc.)*

COMMERCIAL PROCESSES   157

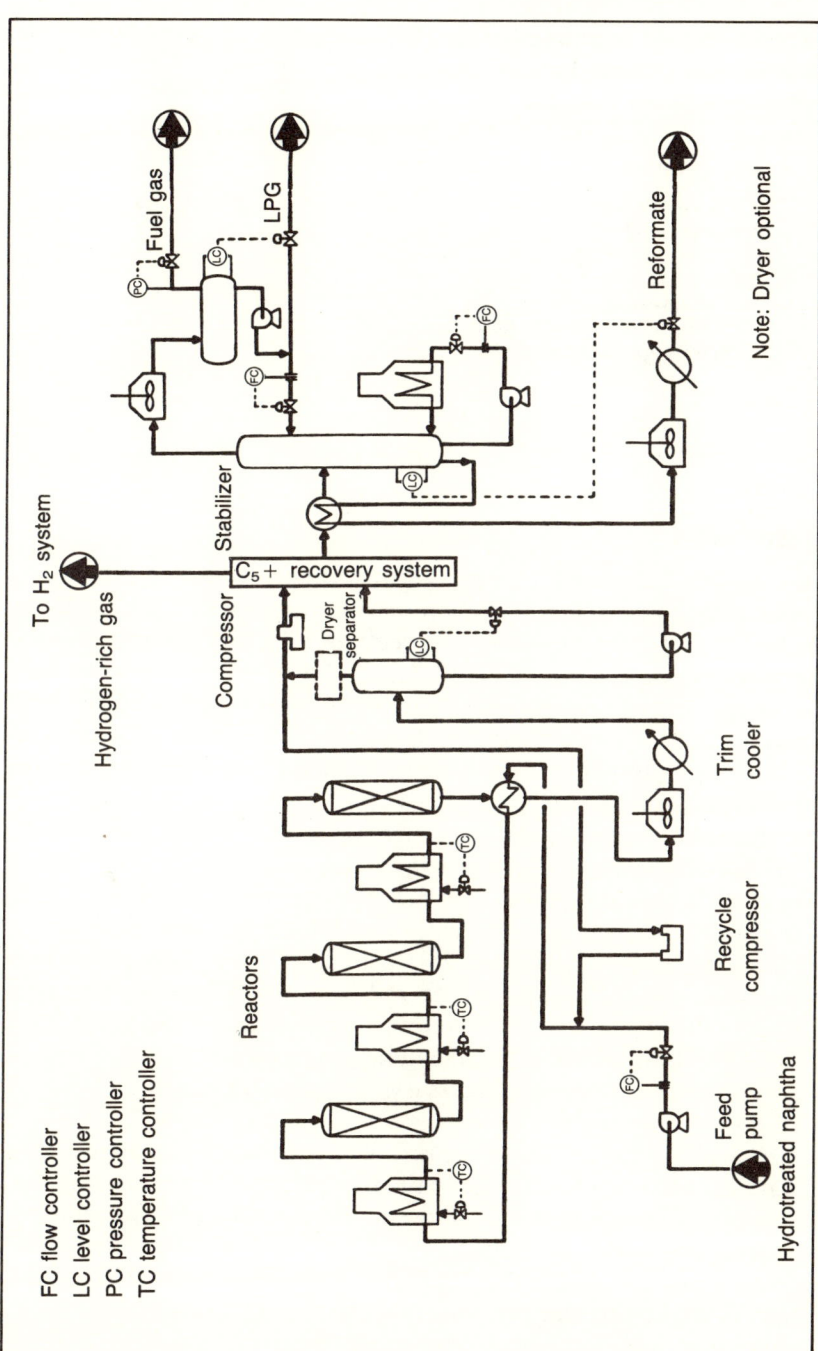

Fig. 7-2 IFP semiregenerative reformer *(courtesy IFP Enterprises Inc.)*

eration circuit is isolated from the reactors at all times by double valves and double bleeder systems, plus an independent system of hardware interlocks.
7. Proprietary catalysts are designed specifically for IFP reforming.
8. Typical utilities per barrel of feed[a]

• Fuel fired, BTU	260,000
• Power, kw-hr	[b]0.57
• Cooling water, gal	[c]143
• Steam, lb	
High-pressure consumption	54
High-pressure generation	58
Medium-pressure generation	54

[a]RON clear of reformate is 100
[b]Hydrogen booster excluded
[c]Delta T = 18°F

MAGNAFORMING

Magnaforming is licensed by Engelhard Industries, Div. of Engelhard Corp. The process design is based on an optimization of catalyst distribution, reactor inlet temperature, gas recycle ratio, and operating pressure. Magnforming design principles are shown in the process flow scheme of Fig. 7–3. The increasing sizes of the reactors from front end to tail end indicate controlled catalyst distribution. The divided hydrogen recycle indicates control of hydrogen-to-hydrocarbon ratio. Significant features of the process are:

1. Controlling (limiting) the temperature and hydrogen partial pressure at the front end of the reactor system to reduce hydrocracking of naphthenes and to increase dehydrogenation to aromatics. The front end reactors are relatively small (high space velocities) and operating temperatures are relatively low. These conditions permit the reactors to operate at reduced H:HC ratios. The first two reactors are held at fixed and ascending temperatures, say 880–900°F throughout a cycle. The temperatures in the terminal reactors are adjusted as necessary to maintain the octane number of the reformate.[3]

2. The hydrogen-rich recycle gas is divided into two streams. One part joins the reactor charge to the first reactor. The other portion, secondary recycle, is added between the second and third reactors. In this mode the hydrogen partial pressure is lower in the two front-end reactors for the benefit of naphthene dehydrogenation and is higher in the last two reactors for the protection of the catalyst at higher temperatures and lower space velocities. Divided

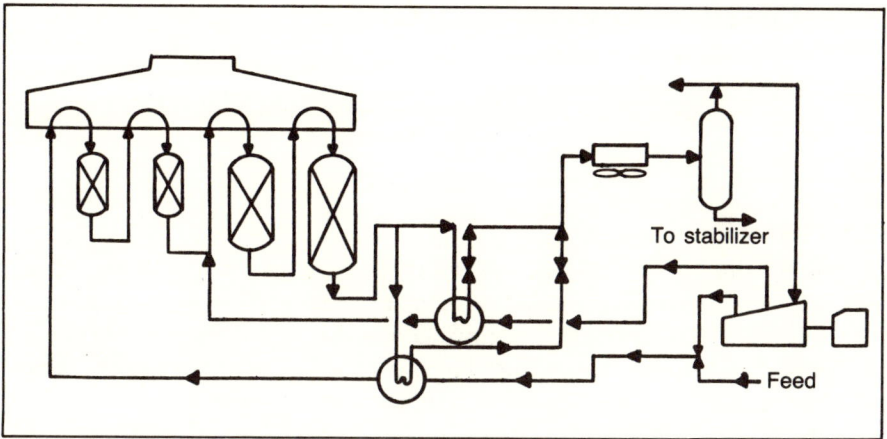

Fig. 7–3 Magnaforming flow diagram *(courtesy Engelhard Industries)*

    recycle gas can be achieved either by two compressors or by a single, side take-off machine.

3. Ascending reactor temperature profile is used with the lowest temperature in the lead reactor and with the highest temperature in the terminal reactor. An increase in reformate yield and catalyst cycle life, for ascending reactor temperature profile and skewed catalyst loading, is reported in a paper presented at the 1971 NPRA annual meeting.[4]
4. Magnaforming (Engelhard) catalysts are specifically designed for use with this process.
5. *Sulfur guard* technology is offered. A single reactor containing a proprietary sulfur adsorbent is placed after the HDS unit or between the first and second reactors of the reformer. The sulfur guard reduces sulfur content of the feedstock to virtually zero. This very low sulfur level gives a two- to three-fold increase in the catalyst stability compared with a feedstock of 1 ppm wt sulfur. The sulfur guard also protects the catalyst from minor upsets in the HDS unit.
6. Typically the process is semiregenerative, but it can be designed for semicyclic regenerative operation.
7. Magnaforming principles can be applied to existing reformers after some revision.
8. Catalyst samplers permit on-stream sampling of catalyst for monitoring catalyst condition.

160 CATALYTIC REFORMING

9. Approximate utilities per barrel of feed[a]

Fuel, BTU	250,000
Power, kw-hr	2
Cooling water, gal	40

[a]125 psig product separator pressure

PLATFORMING

Platforming is licensed by the UOP Process Div. of UOP Inc. This was the first commercial catalytic reforming process to use a platinum catalyst. Platforming was first announced at the 1949 Western Petroleum Refiners Assoc. annual meeting. The process has gone through a series of developments. Older units are still serving as nonregenerative or as semiregenerative reformers. In 1971, UOP introduced their concept of continuous catalyst regeneration as a commercial reforming process. This process is the one we will discuss.

A process flow diagram of the reaction and product recovery sections of a UOP continuous catalyst regeneration unit is shown in Fig. 7–4. Fig. 7–5 illustrates the catalyst regeneration section. The hydrocarbon flow pattern is essentially the same as it is in conventional fixed-bed reformers. Reactor feed combined with recycle hydrogen is raised to reactor temperature by heat exchange and a fired heater and is charged to the reactor section.

Temperature is maintained across the reactor section by interheaters. Effluent from the last reactor is cooled by heat exchange, followed by air and water cooling, and passes to the product-recovery section. The hydrogen-rich separator gas is recycled to the combined feed exchanger and then to the reactors. Excess hydrogen is sent to hydrogen-consuming units or to fuel gas.

In the continuous process, unlike fixed-bed units, the catalyst is continuously withdrawn from the last reactor. The catalyst is continuously regenerated in the regeneration system (Fig. 7–5) and is then returned to the top of the No. 1 reactor (Fig. 7–4). The ability to continuously remove catalyst from the reactor system, restore it to fresh-catalyst condition, and return it to the reactor system permits the unit to operate continuously at high-severity conditions, that is, producing high-octane reformate. A summary of significant features follows:

1. The reactors are stacked one on top of another (Fig. 7–4), which requires less plot plan space with a minimum of piping. This arrangement facilitates the transfer of catalyst from reactor to reactor by gravity flow. Fig. 7–4 shows a three-reactor unit. Reactor internals are generally similar to those of conventional re-

COMMERCIAL PROCESSES 161

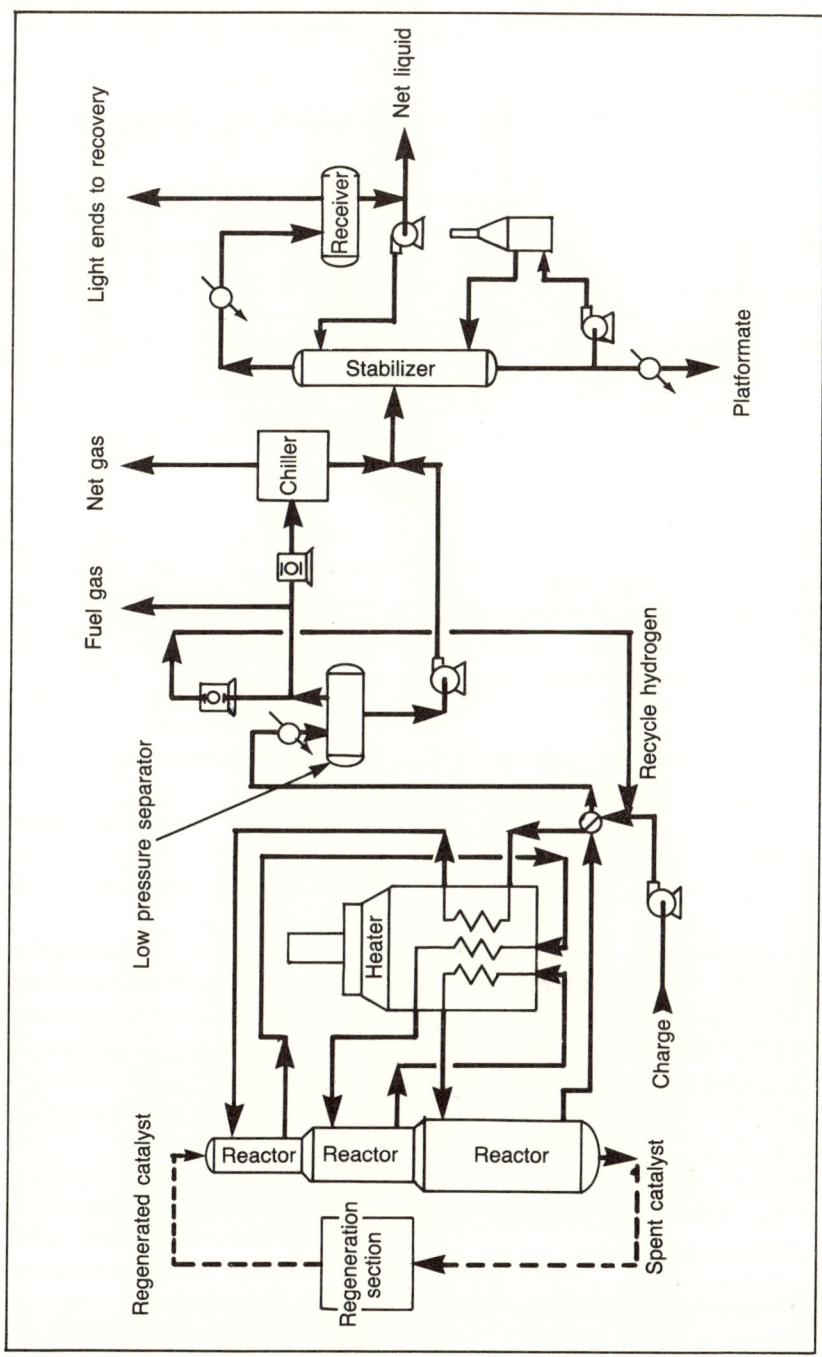

Fig. 7-4 Platforming, UOP continuous catalyst regeneration process *(courtesy UOP Process Div.)*

162 CATALYTIC REFORMING

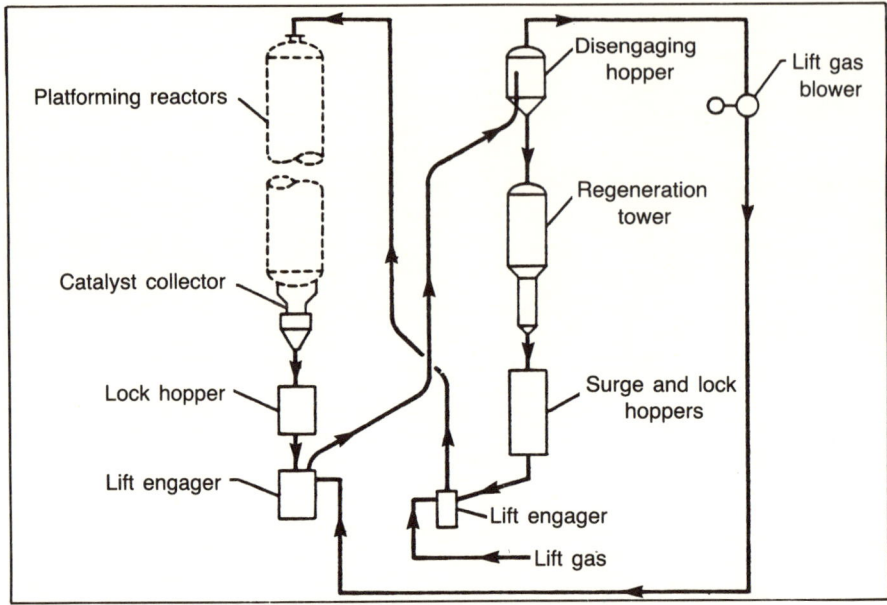

Fig. 7-5 Catalyst regeneration section for UOP continuous catalyst regeneration process *(courtesy UOP Process Div.)*

formers, that is, radial flow with an annular screen and a center pipe. The catalyst moves in essentially plug flow (a minimum of mixing). The rate of flow is determined by the rate of catalyst withdrawal. The combined-feed exchanger, the reactor-section pipe sizes, and the plot plan layout are designed for economically beneficial low-pressure operation.

2. A simplified flow diagram of the continuous catalyst regeneration section is shown in Fig. 7-5. The catalyst regeneration requires five basic operations:
   - Burning off coke
   - Oxidizing metal promoters (e.g., platinum)
   - Adjusting chloride balance
   - Drying off excess moisture
   - Reducing the metal promoters.

   The first four operations occur in the regeneration vessel, and the fifth takes place in the reduction zone on top of the reactor.

3. Safety is a prime consideration in the unit design. The regeneration is monitored and controlled by a system called the Master Controller. The transfer of catalyst between reactor and regener-

ator systems must take place through a sequence of purge and pressure/depressure steps, all of which are automatically controlled.
4. The reactor and regeneration systems can be isolated from one another. The reactor system can continue to operate, leaving the regeneration system available for inspection and maintenance.
5. The reactor system can operate as a semiregenerative unit with in-situ catalyst regeneration. Some units are built this way, with provision for conversion to a continuous catalyst regeneration by addition of the regeneration system.
6. Sampling points provide means for monitoring catalyst condition and for making operating adjustments.
7. Proprietary catalysts designed for specific reforming applications are provided by UOP. The bimetallic catalyst used in continuous-reforming units is attrition-resistant for moving-bed application.
8. Typical utilities per barrel of 100-RON0 feed

Fuel, BTU	314,000
Power, kw-hr	4.5
Cooling water, gal	106

POWERFORMING

Powerforming, the catalytic reforming process licensed by Exxon Research and Engineering, can be designed for either semiregenerative or cyclic reforming. The process flow of Fig. 7–6 shows a semiregenerative scheme and the additional equipment needed for cyclic operation. For most operations, the low-cost semiregenerative process is employed. Powerforming, operating in the cyclic scheme, is the preferred alternative if a very high octane rating or if the maximum yield of aromatics (benzene, toluene, or xylenes) per barrel of feed is desired. *Semicyclic* operations of an existing cyclic unit are anything between cyclic and semiregenerative.

Significant features of the process are:
1. The semiregenerative unit is taken off stream for about a week, at intervals normally varying from 6 to 24 months, while all the catalyst beds are regenerated. Of all modern catalytic reformers, the semiregenerative type requires the least investment and has the lowest operating cost.
2. The cyclic process employs an additional (swing) reactor and a regeneration system that permits regeneration of any individual reactor without taking the unit off stream. Regeneration of one reactor each 24–48 hours is typical, but the frequency can be varied to meet changing process objectives.

164 CATALYTIC REFORMING

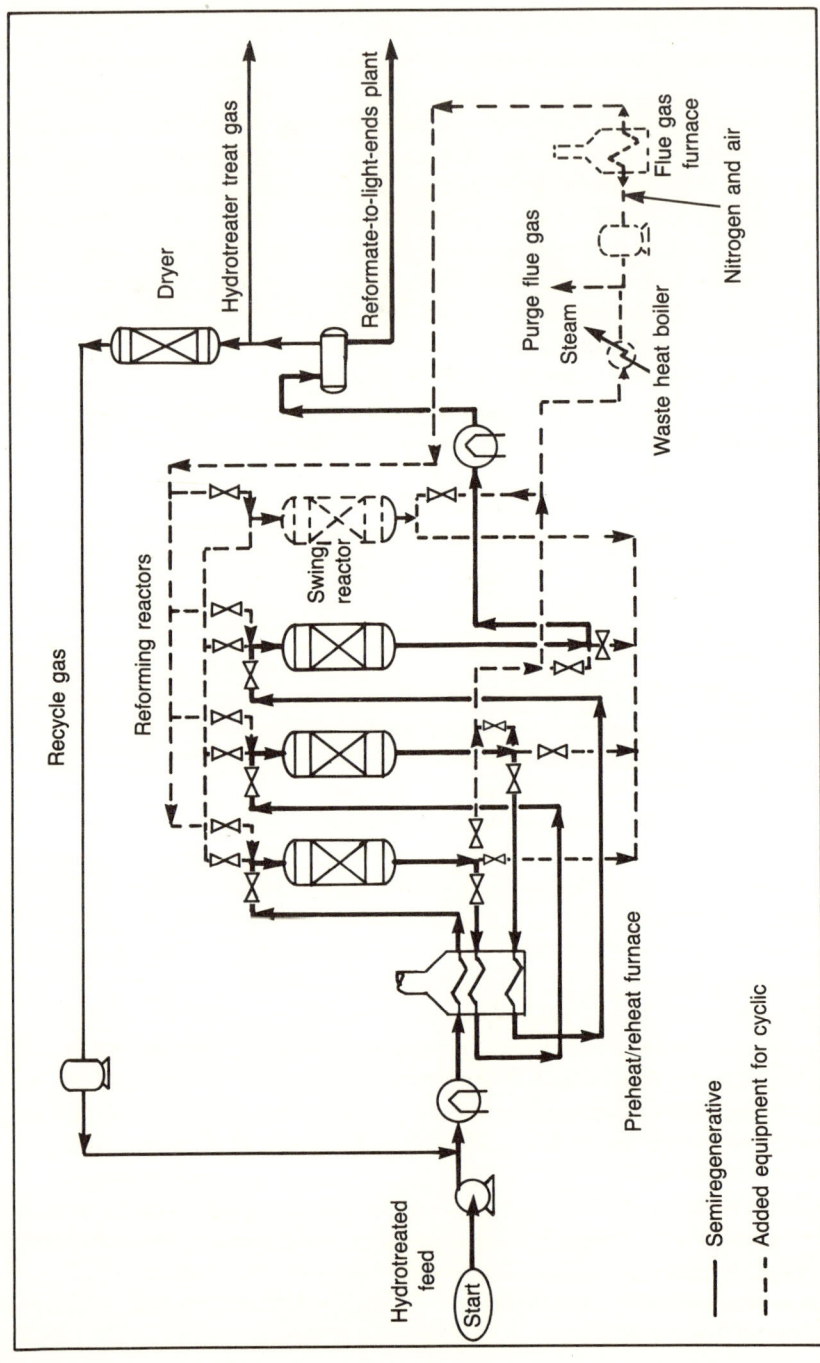

Fig. 7-6 Powerforming unit *(courtesy Exxon Research and Engineering)*

3. Flexibility is inherent in the cyclic design. The swing reactor permits a wide choice of operation between cyclic and semiregenerative. A cyclic unit can operate at very high severity (yielding high-octane-number reformate) without shutting down and, thus, attain long periods of on-stream time.

   The octane number requirement may decrease to the point that cyclic operation is no longer necessary for reasonable cycle time. The regeneration system can then be shut down and the unit operated economically in the semiregenerative mode. The swing reactor operated in parallel to the last reactor saves energy and chemicals.

   Semicyclic operation can be low-severity cyclic operation with increased time between reactor swings. The swing reactor can be switched in to replace the last reactor midway through a cycle, extending cycle length. All reactors can be regenerated while the unit remains on stream.
4. The ability to isolate reactor/furnace pairs is a high-cost option which permits catalyst changeout, major maintenance, and inspection, while the unit continues reforming. This feature is particularly useful when a reactor traps material, causing high pressure drop, or when the catalyst is deactivated by a permanent poison, resulting in shut down of the reactor to replace the catalyst. Many cyclic units have had on-stream runs of 4–6 years. One stayed on stream for over 10 years.
5. Powerformers can be designed and built for initial operation as semiregenerative units, with the provision of later conversion to cyclic units by the addition of a regeneration system.
6. Proprietary Exxon catalysts are used in this process. On cyclic and semiregenerative units, staged catalyst systems (i.e., more than one type of catalyst) are now Exxon's standard. Of course, this multicatalyst system requires compatible regeneration procedures. In a cyclic unit, the catalyst in the swing reactor is chosen to be compatible with all reactor positions (No. 1, 2, 3, or 4).
7. Sulfur control is considered essential to catalyst performance. To fully exploit the advantages of the newer staged catalyst systems, sulfur control is necessary. Powerforming employs a single-vessel-system feed sulfur trap which supplements HDS operations. This trap removes hydrogen sulfide, mercaptans, and thiophenes; it guards against naphtha hydrotreater upsets, as well as further reducing normal feed sulfur levels. Fig. 7–6 does not show a trap, but one would be located between the hydrotreated feed and the first reactor inlet.

8. The purpose of the recycle gas dryer (Fig. 7–6) is to remove moisture and to thus control the moisture in the feed to the reactors. The addition of a hydrogen sulfide absorbent to the desiccant (drying material) prevents recirculation of feed sulfur and of sulfur introduced during catalyst regeneration.
9. Typical utility requirements per barrel of feed are:

Power, kw-hr	2–6
Fuel, 1,000 BTU	200–300
Cooling water, gal	10–70

RHENIFORMING

Rheniforming is a catalytic reforming process licensed by Chevron Research Co. Rheniforming (Fig. 7–7) is a semiregenerative, low-pressure, fixed-bed catalytic reforming process which employs a bimetallic platinum-rhenium catalyst and special operating techniques.

Chevron Research Co. introduced platinum-rhenium reforming to the refining industry in 1967.[5,6] Improvements have since been made in the catalyst, in the process design, in operating procedures, and in regeneration procedures.

Rheniforming features design simplicity, energy efficiency, optimum operation, and a rugged, completely regenerable catalyst. The following illustrates:

1. Rheniformers have few moving parts and no large high-temperature valves or complex manifolding. Newly designed units employ vertical, single-pass, low-pressure-drop heat exchangers and low hydrogen-to-recycle gas ratios to minimize utility requirements.
2. Proprietary operating technology has been developed to assure operation within optimum ranges of the proprietary catalyst and recycle gas properties.
3. Rheniforming catalyst has undergone a number of developments since catalyst A was introduced in 1967. Rheniforming F catalyst, introduced in 1978, has a run length 1.4 times that of catalyst A, despite F's having half the platinum content and a lower density base.[7] Using the catalyst oriented packing technology to load the catalyst into the reactors reduces catalyst-bed settling, reduces maldistribution of flow due to channeling, and contributes to optimum catalyst performance.[8]
4. Chevron's proprietary catalyst samplers permit characterization of the catalyst removed from reactors during a run and indicate necessary operating adjustments to control on-stream catalyst selectivity and activity.

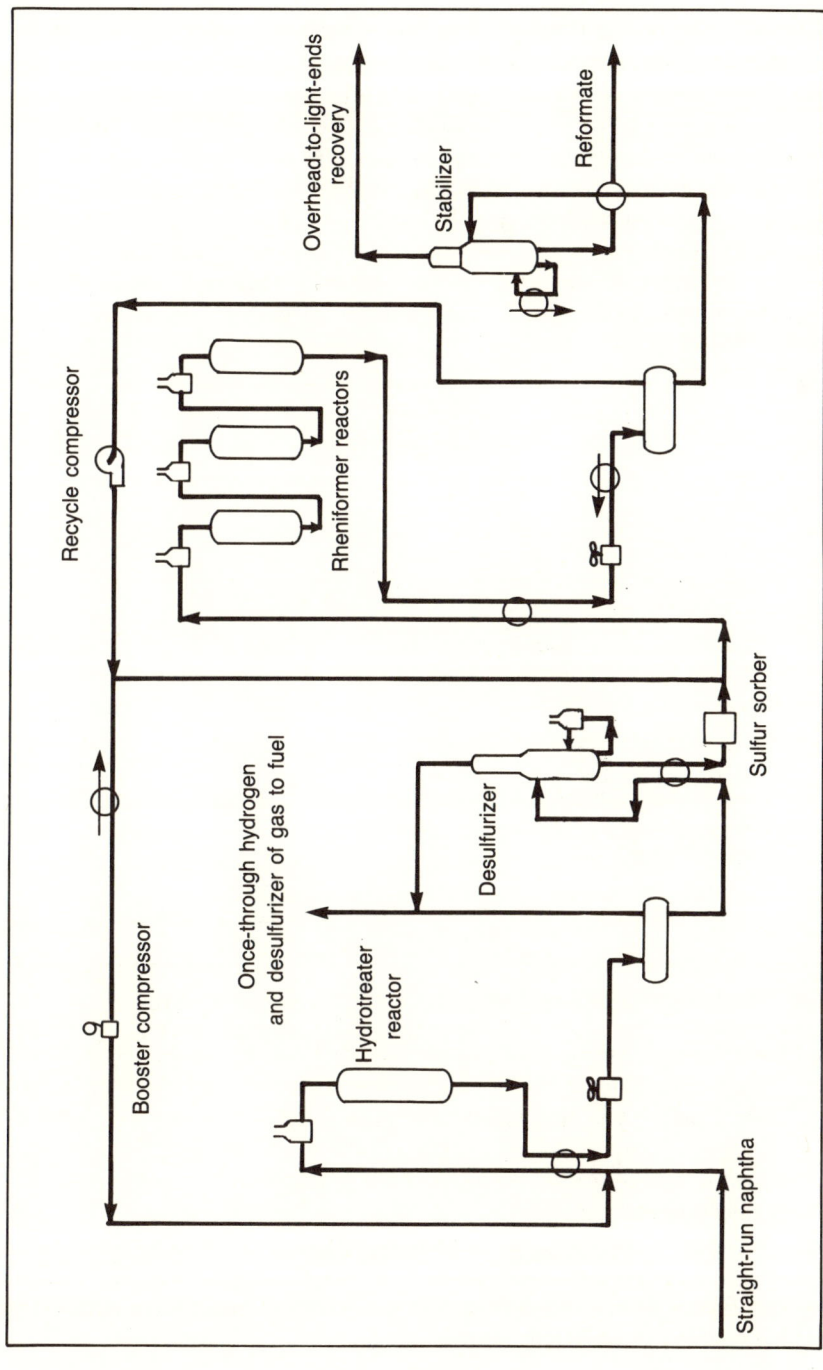

Fig. 7-7 Rheniforming process with hydrotreater *(courtesy Chevron Research Co.)*

168   CATALYTIC REFORMING

5. Regeneration techniques ensure complete recovery of catalyst activity and performance after each regeneration.
6. Rheniforming technology includes sulfur control of reactor feed for optimum catalyst performance. Chevron's information on the relationship between hydrogen sulfide content of the recycle gas and feed sulfur content is shown in Fig. 7–8. Fig. 7–6 shows the Chevron sulfursorber vessel, located between the hydrotreater and the reformer. The sulfursorber contains a proprietary sorbent* catalyst which protects the platinum-rhenium catalyst from hydrotreater upsets and other sources of sulfur contamination.

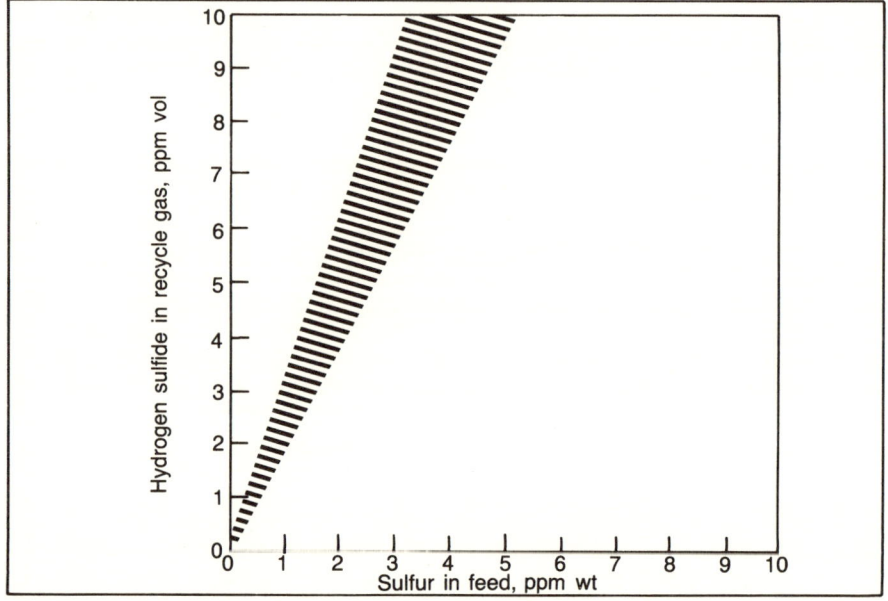

Fig. 7–8 Relationship between hydrogen sulfide content of recycle gas and feed sulfur content *(courtesy Chevron Research Co., NPRA 1980 annual meeting)*

7. Typical utility requirements per barrel of feed[a]

Fuel, gal EFO[b]	2.0
Power, kw-hr	0.7
Cooling water, gal	160

[a]20,000 b/sd Arabian naphtha, 99 RON0 reformate
[b]Equivalent fuel oil

*Sorbent is a catalyst that adsorbs or absorbs.

ULTRAFORMING

Ultraforming is licensed by Standard Oil Co. (Indiana). The technical aspects of licensing are handled by Standard's wholly-owned affiliate, Amoco Oil Co. Ultraforming's first commercial unit went on stream in 1954, utilizing a proprietary platinum-on-alumina catalyst. Starting in the late 1960s Amoco developed and made extensive use of its own proprietary bimetallic (platinum-rhenium) catalyst. Amoco also developed and proved in commercial operation the technology, including regeneration, involved in utilizing this catalyst.

Ultraforming (Fig. 7–9) employs a swing-reactor design, which permits on-stream regeneration for high-severity (high-octane-reformate or high aromatics yields) operation. For lower-octane reformate, the process can be designed as a semiregenerative unit. Significant features are:

1. The design includes a swing reactor, which can replace any of the other reactors during regeneration. Regeneration is carried out while the unit remains on stream, without interrupting process flow or reducing feed rate or octane level. The unit can operate at low pressures and high severity and still remain on stream for long periods.
2. Ultraformers use a proprietary catalyst that has been proven in commercial service. After hundreds of regenerations this catalyst is restored to fresh-catalyst performance level.
3. Amoco's regeneration procedures restore essentially fresh-catalyst activity and selectivity by removing coke deposits, adjusting the catalyst's acid function, and redispersing the active metal components. These procedures, which can be fully automated, are safe and reliable.
4. Safety is an important aspect of the process. All valve operations are sequenced and interlocked to prevent any possibility of error.
5. Ultraformers are designed for operating flexibility. The unit can function in the cyclic or in the semiregenerative mode. The unit can be designed for initial operation as a semiregenerative type and later converted to a cyclic unit by addition of a swing reactor and regeneration system.
6. Ultraforming cyclic units have a vent and block-valve system. This system allows isolation and shutdown of any reactor for regeneration, skimming of catalyst, dumping and screening, and inspection or repair of reactor internals without shutting down the unit.
7. Typical utility requirements per barrel of feed[a]

Fuel, BTU	210,000
Power, kw-hr	4

170 CATALYTIC REFORMING

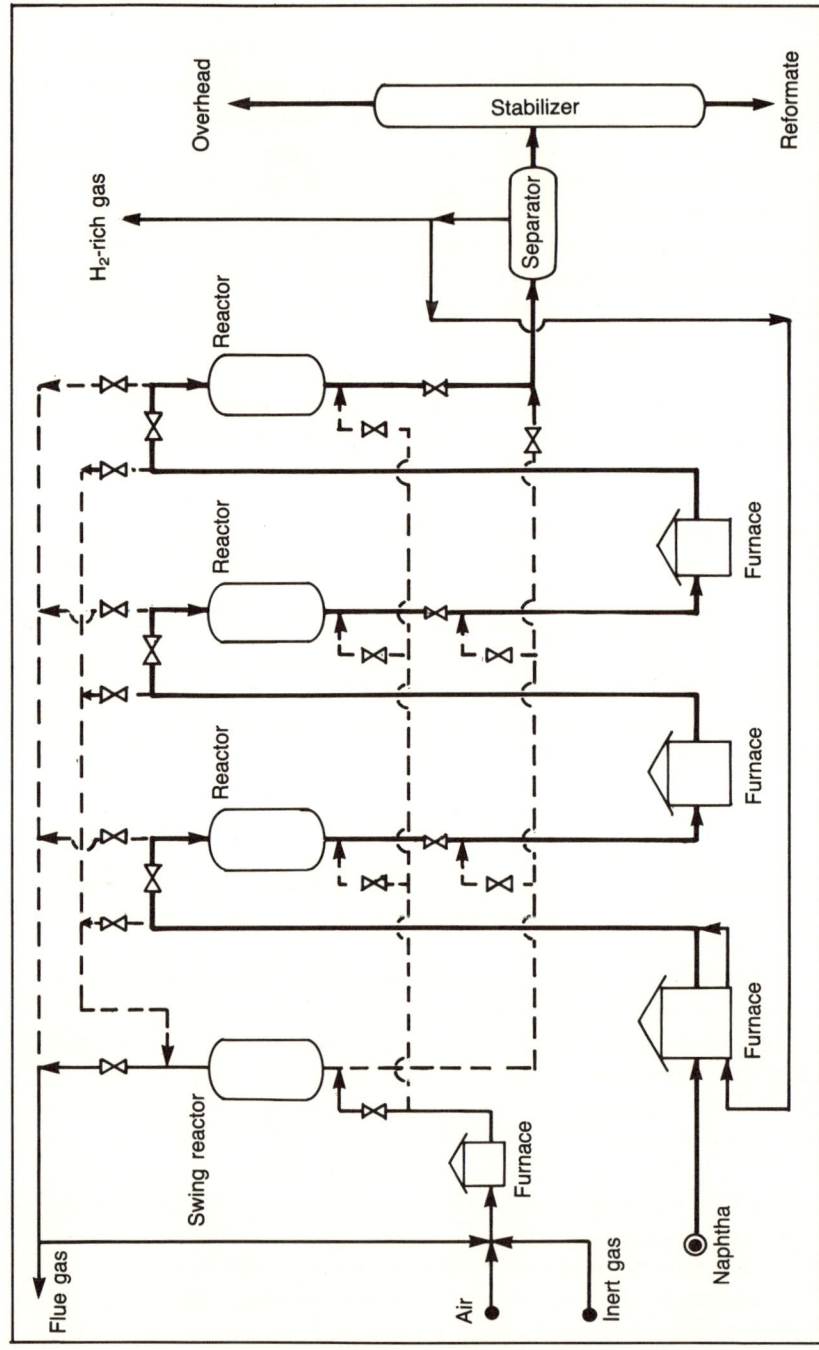

Fig. 7-9 Ultraforming process *(courtesy Standard Oil Co., Indiana)*

Cooling water, gal	Not applicable

^a25,000 b/sd, 100 RON clear ultraformer

STAR

Steam active reforming (STAR) is licensed by Phillips Petroleum Co.; it is not included in Table 7–1 because STAR is not yet in commercial operation. Because it significantly differs in many respects from the other commercial processes in this chapter STAR (Fig. 7–10) is added here. The catalyst is unique. The author followed its development from bench scale through semiworks.

Paraffins containing five or fewer carbon atoms in a chain are dehydrogenated to the respective olefin. Paraffins containing six or more carbon atoms in a chain are dehydrocyclized to aromatics. Other reactions that occur are cracking, which is primarily thermal; a water-gas reaction where the paraffin reacts with the steam that is present to produce carbon dioxide and hydrogen; and, finally, a small production of coke or a tar formation.

These reactions take place in the vapor phase over a fixed catalyst bed using steam dilution. The process is cyclic; while one reactor cell is on regeneration, the others continue on process. Process cycles take 30 minutes, and regeneration cycles last 60 minutes.

The STAR process is characterized by high per-pass conversions and selectivities to aromatics of paraffins from C_6 to C_8. The STAR process includes:

1. A cyclic process for conversion of paraffins from C_6 to C_8 to aromatics, particularly useful for reforming high-paraffinic-content raffinates
2. Heat recovery from reactor effluent is the key to its being an energy-efficient process
3. A proprietary catalyst of unique composition. Hydrogen yield per barrel of feed is typically in the range of 800–1,300 scf/bbl, depending on the feedstock
4. Catalyst is in tubular reactors inside a furnace. Each reactor cell can be isolated and catalyst regenerated while the other reactor cells are on stream
5. STAR operates at moderate pressure and high temperature to yield high-octane-number reformate from high conversion of paraffins to aromatics
6. Its ability to dehydrogenate without isomerization permits STAR to dehydrogenate light hydrocarbons to produce olefins (for example, isobutane to isobutene at high conversion and high selectivity)

172 CATALYTIC REFORMING

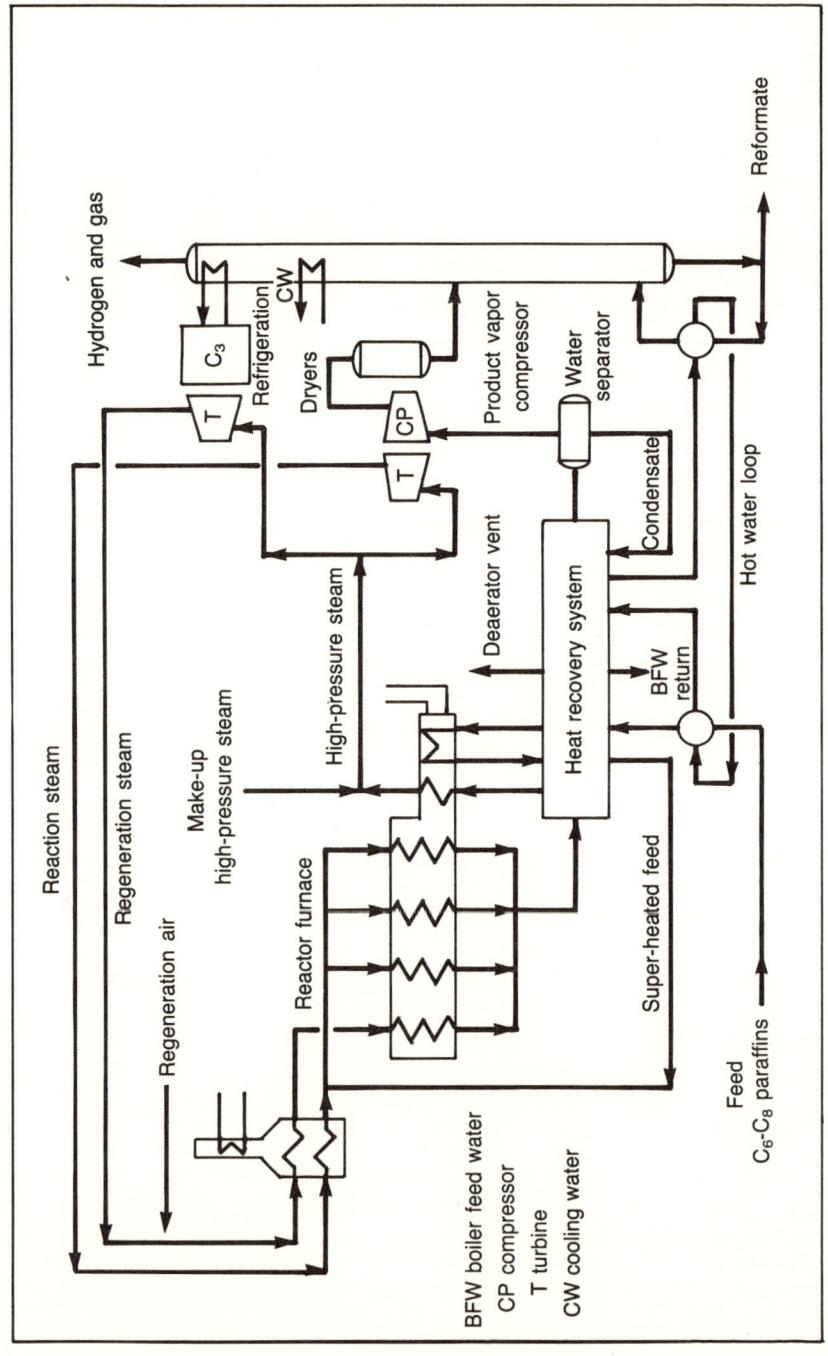

Fig. 7-10 STAR process *(courtesy Phillips Petroleum Co.)*

7. Typical utilities per barrel of feed:

Fuel, 10,000 BTU	55
Power, kw-hr	3.6
Cooling water, gal	1,130
Steam, lb	
600 psig	92
25 psig	22 (Net production)

SUMMARY

The foregoing descriptions of commercially licensed catalytic reforming processes make it clear that anyone contemplating a new reformer or revamping an existing unit has a variety of options. The choice is usually not clear-cut. Anyone familiar with refineries and their differing process configurations realizes that a catalytic reforming process suitable for one refinery may not be right for another. Some general observations about commercial catalytic reforming processes include:

1. The common element of all the above processes is proprietary catalyst. Each licensor has developed a catalyst which he considers to be as good as any on the market, especially if used with his licensed design.
2. Excellent catalysts are available. Some licensors sell their proprietary catalyst for use in a nonlicensed unit. Technical service is provided for using the catalyst. At least one company, American Cyanamid, markets reforming catalysts without requiring a license.
3. Technology—that is, operating and regeneration procedures that best complement a given catalyst—is available from the licensor.
4. Flexibility: catalytic reformer designs are not set in concrete. Once a licensor knows the refiner's reforming needs, adjustments can be made to fit the circumstances, with provisions for estimated future requirements.
5. Estimates can be had for the asking. When supplied with input information, licensors furnish preliminary-yield and cycle-life estimates, catalyst costs, utility requirements, and "ball-park-quality" cost estimates. If cost estimates are promising, further details can often be acquired under secrecy agreements.
6. Others' experience can help. When a process looks good and the estimates come out right, a licensor can arrange a visit to one of its licensee plants for firsthand discussion and observation of an operating unit.
7. Information from licensors is generally reliable because it is based on pilot unit and commercial data. If yields, cycle time, or utilities appear out of line, ask the licensor to confirm the estimate.

REFERENCES

1. HPI Construction Boxscore. *Hydrocarbon Processing.* International Edition, Sec. 2 (February 1984).
2. Refining Developments. *Hydrocarbon Processing* (September 1982): 164–169.
3. Nevison, J.A., C.J. Obaditch, and M.H. Dalson. "Magnaforming with E–601 Catalyst." 72nd annual meeting of the NPRA.
4. Decker, W.H., W.E. Haynes, and M.H. Dalson. "Optimizing Reformer Designs for High Severity Operations." Paper AM–71–6 presented at the annual meeting of the NPRA, 1971.
5. Kluksdahl, H.E. "Reforming a Sulfur-free Naphtha with a Platinum-Rhenium Catalyst." U.S. pat. 3,415,737 (1968).
6. Jacobsen, R.L., H.E. Kluksdahl, C.S. McCoy, and R.W. Davis. "The Chevron Research Rheniforming Process with Platinum-Rhenium Catalyst: A Major New Catalytic Reforming Development." Paper presented at a meeting of the API Refining Div., Chicago, 1969.
7. Freiburger, Marie A., W.C. Buss, and Alan G. Bridge. "Recent Catalyst and Process Improvements in Commercial Rheniforming." Paper AM–80–52 presented at the annual meeting of the NPRA, 1980.
8. Snow, A.I., N.K. Rausch, G.A. Uhl, and L.A. Baillie. "New Catalyst Distribution Method." Paper AM–72–17 presented at the annual meeting of the NPRA, 1972.

CHAPTER 8

Feed Preparation—Naphtha Hydrotreating

A modern 1980s catalytic reformer with bimetallic or multimetallic catalyst in the reactors, producing 90–95 RON clear, or higher, reformate must have properly prepared feedstock to produce maximum yield of reformate and to obtain an acceptable catalyst cycle life. The time has passed when a reformer could operate with feedstock as high as 50–70 ppm wt sulfur and 130–170 mol ppm in the recycle gas.

Current catalysts require less than 0.5 ppm wt sulfur in the feed. The trend is to lower sulfur into the range of 10 ppb wt (chapter 7). Not only sulfur but other contaminants such as water, nitrogen, oxygen, silica, phosphorus, and metals like copper, arsenic, and lead must be removed or reduced to very low levels in reformer reactor charge.

The process by which contaminants are reduced to acceptable concentrations in reformer feed is hydrotreating. The hydrotreating of raw naphtha for reformer feedstock is often called *hydrodesulfurization* (HDS). The processing unit is commonly referred to as the HDS unit because desulfurization is required of nearly all reforming feedstocks. If other contaminants are present, the hydrotreater also denitrogenates, deoxygenates, and demetallates.

The production of hydrogen from catalytic reforming makes hydrotreating an indispensable petroleum refining process. Refiners hydrotreat naphtha, heating oil, jet fuel, diesel fuel, atmospheric gas oil, vacuum gas oil, and residual oil. Sometimes a refiner includes light distillates such as heating oil or jet fuel in the feed to an HDS unit with the naphtha. This discussion is limited to hydrotreating naphtha as a feed preparation for catalytic reforming.

CHEMICAL REACTIONS

The chemical reactions over hydrotreating catalysts are well-known and are illustrated in Table 8–1. These are typical of compounds found in naphtha reforming feedstocks.

The hydrogen consumption of Table 8–1 is a calculated stoichiometric quantity. In actual practice hydrogen consumption is higher. Hydrogen is lost as fuel gas in the stabilizer gas.

Hydrotreating a straight-run naphtha containing 1,500–2,500 ppm wt sulfur and very little nitrogen or oxygen should consume 50–150 scf of hydrogen per barrel of feed. Cracked naphthas, containing 20–30% olefins and diolefins as well as substantial amounts of sulfur and nitrogen compounds will consume 400–800 scf of hydrogen per barrel of feed. These numbers compare with the 800–1,200 scf of hydrogen per barrel of feed produced from catalytic reforming.

Hydrotreating reactions require the breaking of a C—S, C—N, or C—O bond or require saturation of a C═C bond. All of these are exothermic reactions.

In hydrotreating straight-run naphtha, the heat of reaction is mild, and the temperature rise across a reactor is in the range of 0–10°F. For cracked naphtha, rise in temperature will be somewhat higher, 20–40°F.

Reaction rates are much faster, so mercaptans and disulfides are much easier to desulfurize than thiophenes. Gates reports that, at about 660°F, the reaction rate for diethylsulfide is about 15 times that for thiophene.[1] Phenyl mercaptan's reaction rate is about 70 times that of thiophene.

HYDROTREATER DESIGN

Hydrotreaters are very much like catalytic reformers. They are not as complex as reformers because they usually have one reactor instead of three or four. Hydrotreaters may have two reactors, but those are exceptions.

A simplified process flow diagram of a typical hydrotreater is shown in Fig. 8–1. The raw naphtha charge directly from the crude unit has been "cut to 400°F end point," or separated from hydrocarbons which have boiling points above 400°F. The recycle hydrogen, plus makeup hydrogen, joins the raw charge before the charge enters the reactor effluent-to-feed exchanger.

After heat exchange with the reactor effluent, the total charge is raised to the reactor inlet temperature by the charge heater. The total charge of naphtha-plus-hydrogen passes over the catalyst in the reactor. Next the charge passes through heat exchange and cooling and then on to the product separator. Hydrogen, methane, ethane, and butane are

TABLE 8-1
Chemical reactions of naphtha hydrotreating

Desulfurization				Hydrogen Consumed	
				scf/mol[a]	scf/bbl[b]
C—C—C—C—SH n-Butyl Mercaptan C_4H_9SH	+ H_2 →	C—C—C—C + H_2S n-Butane C_4H_{10}			
Mol Wt 90.186 Boiling Point, °F 209				379	3.2
C—C—S—C—C Diethyl Sulfide C_4H_9SH	+ 2H_2 →	2C—C + H_2S Ethane C_2H_6			
Mol Wt 90.186 Boiling Point, °F 198				758	7.0

C═C
| |
C═C
  \\ /
  S
Thiophene
C_4H_4S
	+ 4H_2 →	C—C—C—C + H_2S
		n-Butane
		C_4H_{10}			

Mol Wt 84.138 Boiling Point, °F 183				1,516	12.6

Continued.

178 CATALYTIC REFORMING

TABLE 8–1
Chemical reactions of naphtha hydrotreating—cont'd

Reaction		Hydrogen Consumed	
		scf/mol[a]	scf/bbl[b]
Denitrogenation	Pyridine C_5H_5N + $5H_2$ → C—C—C—C—C + NH_3 (n-Pentane C_5H_{12})	1,895	36.1
Mol Wt	78.1		
Boiling Point, °F	240		
Deoxygenation	Phenol C_6H_5OH + H_2 → Benzene + H_2O (C_6H_6)	379	6.3
Mol Wt	94.1		
Boiling Point, °F	360		
Olefin Saturation	C—C—C—C—C=C + H_2 → C—C—C—C—C—C—C (1-Heptene C_7H_{14} → n-Heptane C_7H_{16})	379	9.5[c]
Mol Wt	98.12		
Boiling Point, °F	200		

[a] Scf hydrogen per mol of compound
[b] Scf hydrogen per barrel of 54°API naphtha containing 1,000 ppm wt of sulfur, nitrogen, or oxygen
[c] For 1.0 vol % olefin in naphtha

FEED PREPARATION—NAPHTHA HYDROTREATING 179

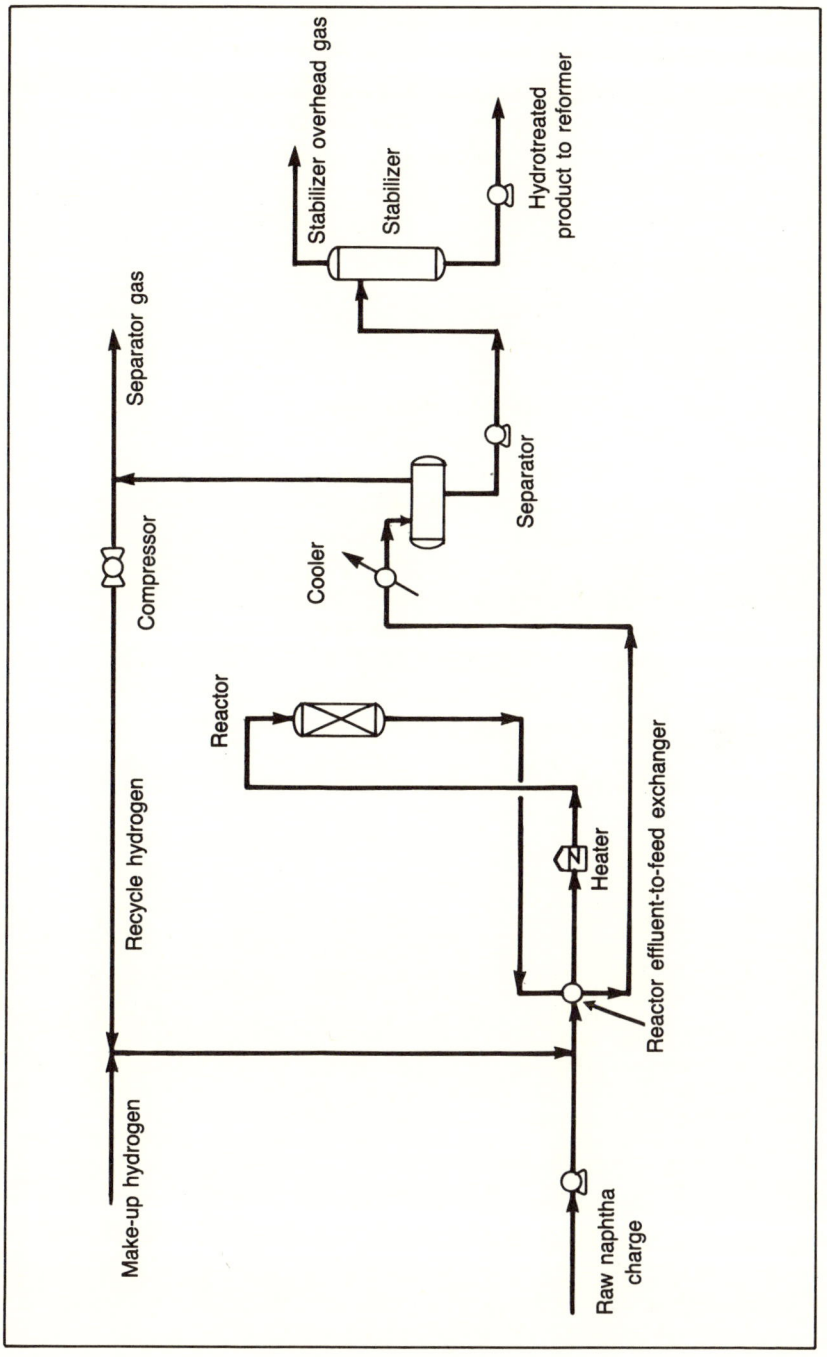

Fig. 8–1 Process flow scheme of a naphtha hydrotreater

180 CATALYTIC REFORMING

separated by flash vaporization and recycled to the reactor effluent-to-feed exchanger. Some separator gas is yielded out of the system to prevent buildup of methane in the recycle gas.

The separator liquid is stripped of hydrogen, light hydrocarbons, hydrogen sulfide, ammonia, and hydrogen chloride in the stabilizer. Stabilizer bottoms are sufficiently free of contaminants to be charged to the catalytic reformer.

OPERATING CONDITIONS

Hydrotreater operating conditions are different from those of reforming. Reactor inlet temperature ranges 525–850°F, and reactor pressure from 400–800 psig. Space velocity is in the range of 2–6 hr^{-1} LHSV, compared with 1–2 hr^{-1} LHSV for reforming.

Hydrogen to the reactor is not expressed as H:HC mol ratio, as in reforming. This hydrogen (recycle plus makeup) is expressed as scf/bbl of feed. The previously mentioned chemical consumption for hydrotreating straight-run naphtha was only about 50–150 scf/bbl. Total hydrogen to the reactor usually ranges 300–600 scf/bbl, for two reasons. The first is to alleviate coking of catalyst; the second is to assure a hydrogen partial pressure above 100 psia in the reactor.

To achieve complete hydrotreating, hydrogen partial pressure should be above 100 psia.[2] Data from commercial hydrotreating of naphtha are listed in Table 8–2. Sulfur in the feed of 1,544 ppm wt is reduced to 0.28 ppm wt in the product, a sulfur removal of 99.98 wt %. Another example (Table 8–3) from pilot unit testing shows catalyst B as more active than A.

TABLE 8–2
Hydrotreating of naphtha

Raw Naphtha Charge, b/sd	21,960
Sulfur, ppm wt	1,544
Hydrotreated Naphtha, b/sd	22,150
Sulfur, ppm wt	0.28
Hydrogen to Reactor, scf/bbl	310
Reactor Inlet, °F	615
Reactor Outlet, °F	615
Reactor Inlet, psig	400
Space Velocity (LHSV), hr^{-1}	4.1

The flow scheme of Fig. 8–1 is one of reactor effluent being cooled and flashed in a single separator. In the interest of energy conservation, some hydrotreaters have two product separators: first, a high-pressure separator which flashes hydrogen recycle; second, a lower-pressure separator which flashes fuel gas.

TABLE 8–3
Pilot unit hydrotreating with commercial catalysts

Feed Properties			
API gravity	64°		
Sulfur, ppm	350		
Nitrogen, ppm	3		
Pressure, psig	190		
LHSV, hr^{-1}	2.5		
H$_2$:oil, scf hydrogen/bbl feed	375		
Product Sulfur, ppm			
	500°F	550°F	600°F
Catalyst A	1.5	1.0	0.5
Catalyst B	<0.5	<0.5	<0.5

(Courtesy Katalco Corp.)

The HDS reactor can be downflow or radial flow. Most are downflow and are designed like the reformer downflow reactor of Fig. 3–6.

It is important that HDS reactors have the trash-collector baskets in the top of the catalyst bed. HDS reactors are well-known for fouling or crusting at the top of the catalyst bed because of scale, polymer, and fines.[3]

A newly loaded downflow reactor with fresh catalyst should have a pressure differential of 5–15 psi. Fouling and crusting at the top of the bed causes pressure buildup until the differential is 30–50 psi. The reactor is usually shut down because inlet pressure becomes so high that hydrogen flow to the reactor is restricted. The reactor is opened, and the top of the bed is skimmed to remove crust and scale. Fresh catalyst is added to replace that which has been removed by skimming. The unit is then placed back in service.

HYDROTREATING CATALYST

Hydrotreating catalysts are almost always cobalt-moly, nickel-moly, or cobalt-nickel-moly, on an alumina base. There are a few nickel-tungsten and a few that may include palladium for special applications. A naphtha HDS in front of a catalytic reformer normally uses a cobalt-moly or nickel-moly or cobalt-nickel-moly catalyst.

HDS catalysts are cylindrical extrudates of 1/8-in., 1/10-in., 1/16-in., or 1/20-in. diameter. These are also available in shapes of trilobes, quadralobes, spheres, or rings.

The compositions of hydrotreating catalysts is common knowledge in the industry (see Table 8–4). Catalyst manufacturers' technical reports and bulletins usually include chemical composition along with other properties of catalysts.

182 CATALYTIC REFORMING

TABLE 8-4
Composition of hydrotreating catalysts[a]

MoO_3, wt %	13–20
CoO, wt %	3–6
NiO, wt %	3–5
Average Bulk Density, lb/cu ft	35–45
Surface Area, sq m/gm	150–250

[a] Based on general information available in technical bulletins (catalyst may be Co-Mo, Ni-Mo, or Co-Ni-Mo)

Hydrotreating catalysts deactivate slowly due to coke laydown. Reactor inlet temperature is raised to maintain desulfurization. The catalyst can be regenerated by oxidation of the carbon on the catalyst. Carbon is oxidized with air, using either steam or nitrogen diluent. The catalyst loses 5–15% activity each regeneration; after two or three generations the catalyst is replaced with new catalyst. There is no known method for rejuvenating HDS catalyst.

Some refiners use a *standard activity test*. A sample of spent catalyst is laboratory regenerated, and its performance is compared with that of the new catalyst. In general, after laboratory regeneration a catalyst with an activity no less than 80% of new catalyst should be regenerated and used again. A catalyst with an activity lower than 80% of fresh should be discarded. HDS catalysts usually achieve catalyst lives of 300–700 bbl/lb before being discarded.

A standard activity test is useful for comparing HDS catalyst from different suppliers. The feedstock in HDS catalyst tests is nearly always a virgin gas oil or a cat-cracked cycle oil instead of naphtha. Using the heavier oil as standard feedstock gives more information because a catalyst that can desulfurize the heavier feed should easily desulfurize lower-boiling feedstocks. If a refiner does not have access to laboratory testing, most catalyst suppliers will run their standard activity test free.

Standard activity tests on two different catalysts are shown in Fig. 8–2. Catalyst A is a nickel-moly catalyst, removed from a commercial naphtha HDS reactor after 197 bbl/lb feed cycle life. This catalyst had been regenerated once before and was reloaded for a second cycle. According to the test, this catalyst, after its second regeneration, desulfurized to the same level of sulfur in the product at a temperature 15°F higher than expected of new catalyst.

Catalyst A could be used again, after regeneration. The cobalt-moly catalyst B, in the same standard activity test, showed a 30–35°F higher temperature requirement for reducing sulfur in the product to new-catalyst level. This catalyst was judged as not good enough for another cycle and was discarded.

FEED PREPARATION—NAPHTHA HYDROTREATING

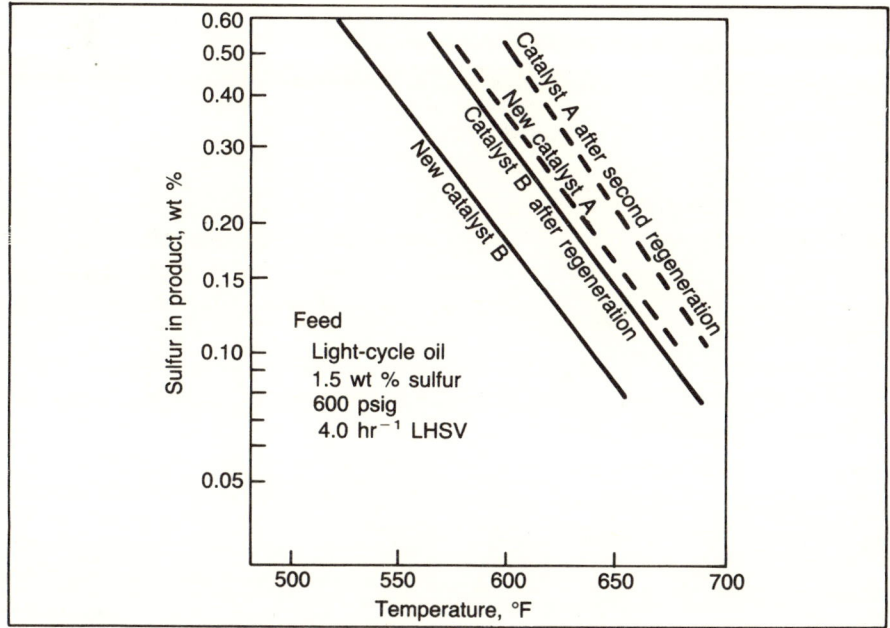

Fig. 8-2 Standard hydrodesulfurization activity test

Merchant regeneration is widely used for hydrotreating catalyst. The catalyst is returned to the refiner with carbon removed, free of fines, and ready to load into a reactor.

Hydrotreating catalyst does not contain precious metals. So the cost is much lower than that of reforming catalyst. Most refiners have a spare load or two of regenerated HDS catalyst on hand. If the HDS catalyst is proven capable, a refiner has the option of using the same HDS catalyst in a distillate or gas oil hydrotreater as well as in a naphtha HDS unit.

The choice between a cobalt-moly and a nickel-moly catalyst for hydrotreating is more or less plant preference. Both types reduce sulfur in the HDS product to that required for a reformer feedstock. The nickel-moly catalyst requires a temperature 10–20°F higher than that of cobalt-moly for the same desulfurization efficiency. However, nickel-moly is a better denitrogenation catalyst than cobalt-moly.

If nitrogen in the raw naphtha is 5 ppm wt or higher, nickel-moly catalyst would probably be chosen. As a rough approximation, a catalyst with 99% desulfurization activity would be expected to have a denitrogenation activity in the range of 60–80%. If olefins, oxygen, or polymer-forming contaminants are in the raw naphtha, a nickel-moly catalyst generally shows higher activity than a cobalt-moly one.

184 CATALYTIC REFORMING

Unless poisoned by heavy metals, hydrotreating catalysts in naphtha service have long lives. The figure varies with the feedstock and severity, but many refiners report over a 1,000-bbl/lb life.[4]

METALS ADSORPTION ON CATALYST

In a metals-poisoning environment (see chapter 5), refiners rely on the hydrotreating catalyst to act as a guard for the reforming catalyst. Arsenic, lead, copper, and the like are poisons and deactivate hydrotreating catalyst, but refiners consider it better to sacrifice a cobalt-moly or a nickel-moly catalyst than to poison a bimetallic reforming catalyst.

The capacity of HDS catalyst for heavy metals was discussed at NPRA Q&A sessions.[5] The following comments are based on the author's observation:

- Both arsenic and lead chromatograph through a bed of HDS catalyst.
- Cobalt-moly or nickel-moly catalyst can hold at least 0.5 wt % total arsenic-plus-lead before a breakthrough of sufficient quantity to rapidly deactivate the catalyst in a reformer downstream of the HDS unit. This means less than 20 weight parts per billion (ppb wt) of arsenic-plus-lead is in the reformer charge, though some put the limit at 10 ppb wt.
- Cobalt-moly or nickel-moly HDS catalyst holds at least 2 wt % arsenic-plus-lead, before deactivating to the point of being spent for HDS operation.
- The HDS reactor must operate above a minimum temperature to break down metal compounds.[6] This temperature is in the range of 525–575°F. If metal breakthrough occurs at temperatures this low, then raising reactor temperature may reduce the metal content of the reactor effluent.
- A good method for removing lead (but not arsenic) from hydrocarbon is the use of silica gel.[7] A plant with a large volume of TEL-contaminated naphtha used this process to successfully clean up reformer feedstock.

Table 8–5 demonstrates the effect of metal contamination on HDS catalyst activity. The cobalt-moly catalyst was in naphtha HDS service six years without regeneration. As the catalyst was unloaded, samples approximating the bottom, middle, and top of the bed were taken as the catalyst flowed out the reactor-bottom dump nozzle.

Two items from Table 8–5 are of particular interest. First, the arsenic was not removed from the catalyst in a laboratory regeneration. Second,

TABLE 8-5
**Metal contamination and deactivation
of cobalt-molydenum HDS catalyst**

Catalyst Bed Location[a]	Top	Middle	Bottom	
As Received				
Carbon, wt %	2.9	6.5	3.4	
Sulfur, wt %	3.6	3.0	4.6	
Arsenic, wt %	0.61	2.7	0.014	
Lead, wt %	NA[b]	1.0	NA[b]	
After Regeneration	0.61	3.6	0.01	
Arsenic, wt %				
Standard HDS Catalyst Activity Test[c] After Regeneration				New Catalyst
Sulfur in Product at 650°F Reactor Temperature, wt %	0.17	1.06	0.12	0.16

[a]Approximate bed location. Samples were obtained from reactor-bottom nozzle as catalyst was unloaded.
[b]Not analyzed.
[c]Standard activity test same as in Fig. 8-2.

the catalyst with 0.61 wt % (top sample) and 0.014 wt % (bottom sample) arsenic tested the same activity for desulfurization as for new catalyst, but the catalyst with over 3 wt % (middle sample) arsenic showed very poor desulfurization.

SULFIDING HDS CATALYST

Hydrotreating catalysts used in the 1980s must be sulfided (treated with sulfur) before their initial use. After regeneration they are not usually sulfided. New HDS catalysts are sulfided to form the sulfur-metals complex which activates the cobalt-moly and nickel-moly catalysts. In addition, sulfiding attenuates superactive catalyst sites which would otherwise promote hydrocracking and shorten cycle life. Sulfiding is carried out by passing hydrogen and a sulfiding agent over the catalyst at a temperature no higher than 450°F until hydrogen sulfide breaks through.

The sulfiding agent is usually a virgin distillate containing at least 1 wt % sulfur. Sometimes a refiner will inject a sulfiding agent such as ethyl mercaptan, butyl mercaptan, dimethylsulfide, or dimethyldisulfide. The mercaptan breaks down to hydrogen sulfide more readily than the sulfide compounds, but refiners differ in their preferences.[8]

Completely sulfiding HDS catalyst takes about 10 wt % sulfur on the catalyst. Cracked stocks are not recommended for sulfiding HDS catalyst—coking and polymer formation interfere with proper sulfiding.

MONITORING NAPHTHA HYDROTREATING

Monitoring naphtha hydrotreater operation essentially means testing the hydrotreated product each day to be sure sulfur is being removed to specified limits from the raw naphtha. Sulfur can be checked daily, but sometimes reformer operation is the first signal of a sulfur breakthrough on the hydrotreater. The reformer catalyst responds to sulfur poison within a few minutes. Analysis for metals such as arsenic and lead is infrequent, unless the reformer catalyst shows symptoms of metals poisoning.

Reactor inlet temperature is the process variable for monitoring catalyst activity. An inlet-temperature increase signals a catalyst-activity decrease unless sulfur in the reactor charge has increased. Data from the naphtha HDS unit are included in the material balance of Table 5–9.

TROUBLESHOOTING AN HDS UNIT

Troubleshooting a hydrotreater is, in principle, the same as troubleshooting other operating units. The engineer must be aware of sudden and subtle changes that signal trouble and must recommend corrective action. A common problem on HDS units is an increase in sulfur content of the product. A number of things that could cause increased sulfur are:

- Increased sulfur in feed or more thiophene in the feed
- Leaks in the reactor effluent-to-feed exchanger
- Inadequate stripping of hydrogen sulfide in the HDS stabilizer
- Insufficient hydrogen in the reactor
- Poor flow distribution (channeling) in the catalyst bed
- Recombination of sulfur with olefin
- Coking of the catalyst.

Too much sulfur in the product is frequently caused by the leaking of feed into the reactor effluent, in the effluent-to-feed exchanger. This possibility should be one of the first investigated.

Another possible source of high sulfur is poor stripping in the HDS stabilizer. This condition can be easily determined by having the laboratory test for hydrogen sulfide in the stabilizers' bottoms. The laboratory can also test for higher-than-normal sulfur in the raw naphtha. In this case, raising the reactor temperature should reduce sulfur in the product.

Testing for a change in type of sulfur in the feed is difficult and time consuming. Investigate this possibility only after others are eliminated.

With a reformer down, a refinery can have an HDS reactor short of hydrogen. At such times, the HDS reactor can operate with less than

100 psia partial pressure of hydrogen. A possible result of this low pressure is higher sulfur in the product.

Hydrogen has been consumed before reaching the bottom of the catalyst bed. When this happens, there is insufficient hydrogen to desulfurize the feed in the lower part of the bed.

Another circumstance that can rob a reactor of hydrogen is high pressure-differential across the reactor. Reactor inlet pressure increases when this occurs. The hydrogen makeup compressor or the recycle compressor may not work efficiently against the higher pressure, resulting in a decrease in the quantity of hydrogen entering the reactor.

Poor flow distribution, or channeling, through the catalyst bed raises space velocity. Desulfurization may be incomplete. Raising temperature does not help much in this situation; it only increases the rate of coking, aggravating the problem.

RECOMBINATION OF HYDROGEN SULFIDE

One of the most puzzling occurrences in HDS units is the recombination of hydrogen sulfide with olefins, either in the reactor or beyond the reactor's outlet. This recombination is an established phenomenon, reported by a number of hydrotreating technologists.[9,10,11] The occurrence is associated with olefins from cracked naphtha feedstocks or from high-temperature hydrocracking in the reactors.

An example of a commercial experience reported by Mauleon[12] follows:

Reactor Inlet, °F	Sulfur in Product, ppm wt
610	1.1
600	0.5
590	0.3
585	0.1–0.2

Tummermans reported chemical equilibrium sulfur levels in the product can be calculated from the bromine number of the product, the partial pressure of the hydrogen sulfide, and the temperature of the hydrogen-hydrocarbons mixture.[13] He presented the following tabulation:

Bromine Number of Product	1.0	2.0	4.0
	Equilibrium Sulfur, ppm wt		
	Partial Pressure of Hydrogen Sulfide (kg/sq cm)		
Temperature, °C	Sulfur in Product, ppm		
300	0.6	1.2	2.5
375	2.0	4.0	8.0
400	1.5	3.0	4.0

Thus, at 375°C, bromine number of 2.0, and partial hydrogen sulfide pressure of 0.25 kg/sq cm, an equilibrium sulfur of 1 ppm wt is reached: $0.25 \times 4 = 1$. At a bromine number of 4.0, the equilibrium sulfur is $0.25 \times 8 = 2$ ppm wt.

Recombination of hydrogen sulfide and olefins is associated with high-temperature hydrocracking. Ordinarily, when too much sulfur is in the product, there is a tendency to raise reactor inlet temperature. However, if recombination is the cause of this problem, lowering reactor temperature may be the solution.

The procedure is to lower reactor inlet temperature at least 50°F—in 10°F increments—and to observe the results. If there is no improvement, then try raising the temperature.

COKING

Carbon laydown, or coking, of the catalyst covers active sites, so temperatures must be raised. Coking is a gradual process. Sudden coking of catalyst can be a result of loss of hydrogen or of inclusion of large amounts of high-boiling hydrocarbons in the feed.

NITROGEN

Sudden increase in nitrogen in the HDS product will be detected on the reformer. Analysis for nitrogen is not regularly scheduled for HDS units. Once nitrogen is suspected, the raw naphtha charge and the HDS product should be analyzed for nitrogen content. If the nitrogen in the raw naphtha is higher than past levels, then a search for the sources is in order. Search for sources upstream of the HDS unit. Most likely, the sources are amine inhibitors or antifouling agents that got into the raw-naphtha charge. The same procedure applies to oxygenated hydrocarbons, arsenic, and lead.

SILICA

A contaminant not yet mentioned, but one that gives refiners plenty of trouble is silica.[14,15] The most common silica source is antifoam carried in coker naphtha. Another silica source is antifoam used in fractionators. The polysiloxane antifoam agent breaks down on contact and deposits silica on the catalyst.

A suggested remedy is to reduce the amount of antifoam used in cokers and fractionators. Another suggested remedy is the use on top of the catalyst bed of support balls containing a low concentration of nickel-moly. The activity imparted to the support balls facilitates antifoam-agent breakdown and deposits silica on the support instead of on the catalyst.

DEPOSITS ON CATALYST

Silica is only one of many materials known to foul the top of the catalyst bed in HDS reactors. Catalyst fines, scale, coke, polymer, and caustic are some materials that deposit in the top layer of the catalyst. They result in poor flow distribution and an increase in pressure differential across the reactor. The fouling can be either blockage by fines in spaces between catalyst particles or crusting of the top layer of catalyst.[16]

After a time, the reactor must be opened and the top layer of the catalyst bed removed by skimming. Since skimming requires entry into the top of the reactor, inert entry must be made under a nitrogen atmosphere.[17] The catalyst is broken up and vacuumed off to remove the crusted layer. Polymers are deposited, not only on top of the catalyst but also in the exchangers and in the feed heater tubes.[18]

In chapter 5, it was pointed out that dissolved oxygen is the major cause of polymer fouling.[19] Preventive measures practiced by refiners are gas-blanketing feed tanks, use of floating-roof tanks, direct rundown of feedstocks from the crude unit, and use of antifoulants. Such methods alleviate polymer deposits but seldom eliminate them.

Another cause of deposits, particularly in feed heaters, is passing through the dry point, at which the naphtha feed becomes completely vaporized.[20] This is the place where deposits frequently occur. Such deposits in heater tubes reduce heat transfer rate. The high tube-skin temperature necessary to compensate for loss of heat transfer rate can lead to tube failure and to fires. Some refiners make certain the naphtha is completely vaporized by heat exchange before the feed enters the fired heater. This moves the deposits from the heater to the exchangers, but refiners would rather clean an exchanger than have a fire in a heater. One HDS unit that suffered repeated heater-tube failures solved the problem by deoiling the naphtha, thus removing a high-boiling polymer-forming material.

CHLORIDE DEPOSIT AND CORROSION

Ammonium chloride deposits and chloride corrosion get a lot of attention from refiners.[21] The source of chloride is organic chloride in the raw-naphtha charge and hydrogen chloride in hydrogen from the reformer.

As discussed in chapter 5, chloride deposits are controlled by water washing or by filming amine inhibitors. Control of hydrogen chloride corrosion is by ammonia or by inhibitors, or both. The best control method is elimination of the chloride source. Each HDS unit differs

190 CATALYTIC REFORMING

from other HDS units; a control method that works for one may not work for another.

For example, one unit was successful using water wash with ammonia injection. Anhydrous ammonia was the better corrosion-control method used on another unit. A third unit found a filming amine sufficient for control.

The quantity of chloride in the raw-naphtha feed is important. If chloride in the raw-naphtha feed is 7 ppm wt or less, control with ammonia is usually no problem. Chloride in the range of 10–20 ppm wt makes corrosion control very difficult. At 25 ppm wt chloride or higher, the HDS units I have monitored have corroded, regardless of attempts to control the corrosion.

A study which developed a method for corrosion control at high chloride concentrations revealed no catalysts nor adsorbents that would remove organic chloride from naphtha. However, organic chloride is converted to hydrogen chloride in the HDS reactor. There are adsorbents on the market that adsorb hydrogen chloride from the reactor effluent. The adsorption capacity is about 1 lb chloride per pound adsorbent.

Size 13x molecular sieves remove hydrogen chloride from the reactor effluent, with a capacity of about 2 lb chloride per pound molecular sieve. The sieves are destroyed in this service and must be discarded.

In both cases, the cost of adsorbent was prohibitive. Articles on the subjects of fouling and corrosion in HDS units by chloride and by ammonium sulfides are of interest.[22,23,24]

SAFETY

The discussion in chapter 5 concerning hydrogen at high pressure applies to hydrotreaters as well as to reformers. In the case of naphtha hydrotreaters, the possibility of a reactor temperature excursion is not considered as likely as in reformer reactors. There are no known naphtha HDS reactors with skin thermocouples on the shells. For reasons previously explained, HDS units are more likely to have heater-tube failures than are reformers.

Since the heaters are upstream of the reactor, the sudden release of pressure at the heater reverses normal flow. The hydrogen and hydrocarbon in the reactor and in the product separator backflow into the heater firebox, adding fuel to the fire. Therefore, most HDS units and some reformers are equipped with quick-opening valves on the reactor that permit rapid relieving of hydrogen and hydrocarbons to the flare, depressuring the system.

Another safety hazard is possible formation of nickel carbonyl when nickel-containing catalyst is in the reactors. Nickel carbonyl, $Ni(CO)_4$, is an extremely toxic, colorless gas which is formed by the contact of carbon monoxide and nickel in the reduced state.[25] The Occupational Safety and Health Administration (OSHA) permissible exposure limit is 1.00 ppb. Defining safety procedures is not my purpose here. Each supplier of a nickel-bearing catalyst has instructions which should be strictly followed.

A second hazard, mentioned with respect to the nickel-containing hydrotreating catalysts, is nickel subsulfide, Ni_3S_2. This compound is reportly a carcinogen.[26]

Each user of a nickel-bearing catalyst should seek the advice of his safety personnel. Aside from the two safety hazards mentioned, the usual precautions in handling dusty, sulfided catalyst, in accordance with plant procedures, should be observed.

CATALYST SUPPORT MATERIALS

A review of the manner in which catalyst is supported in commercial reactors revealed that there is almost no published information on this topic. In most refineries, catalyst support balls are loaded in a certain way because "that is the way it has always been done." Catalyst suppliers usually offer suggestions. Some guidelines are:

1. Support materials (balls, pellets, or granules) are placed in a reactor, both below and above the catalyst.
2. The objective is to support the catalyst in the bottom of the reactor. The support is sized to prevent migration to the reactor outlet screen, where the catalyst might plug the screen. But the support must not restrict vapor flow to the point that pressure drops excessively.
3. The most widely practiced rule seems to be that a support must not be greater than twice the diameter of the material with which it interfaces. For example, 1/16-in. catalyst extrudates rest on 1/8-in. support balls which rest on 1/4-in. balls which rest on 1/2-in. balls which rest on 1-in balls (Fig. 8–3). However, it is common for 1/2-in. balls to rest on 3/4-in. balls instead of on 1-in balls. There are exceptions; for example, 1/20-in. extrudates rest on 1/8-in. balls with no apparent problems.
4. The depth of each layer of support varies but is in the range of 3–8 in. After the bottom outlet nozzle is covered with support balls, 3-in. layers of each size are sufficient. A catalyst dump nozzle at the bottom of a reactor is usually filled with support materials.

192   CATALYTIC REFORMING

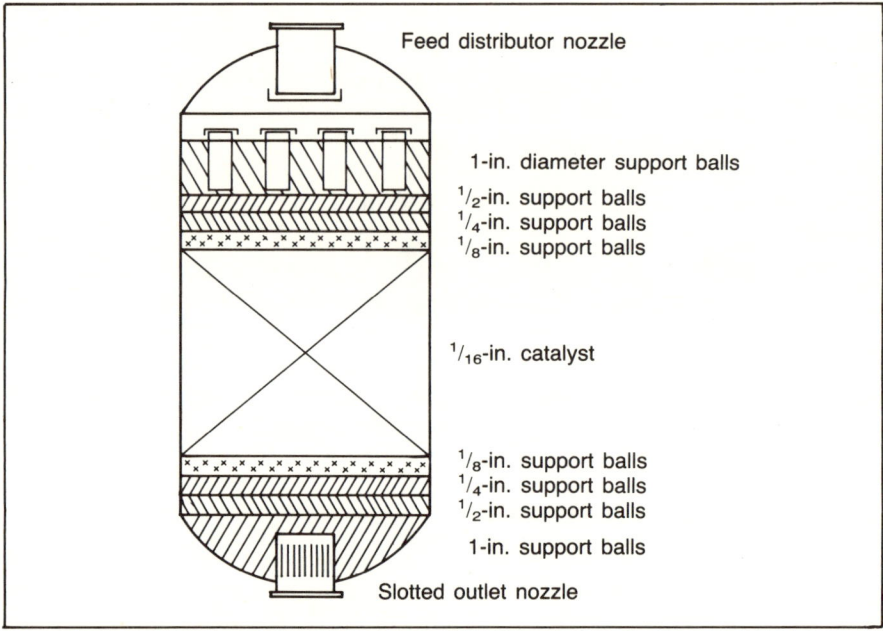

Fig. 8–3 Catalyst support scheme

5. Support materials are placed on top of the catalyst bed. Gradations are the reverse of those at the bottom of the bed. Support materials are placed on top of the catalyst to distribute incoming vapors so there is even distribution of flow through the catalyst bed.

In addition, the support materials on top of the catalyst protect the top layers of the catalyst from impact of incoming high-velocity vapors, which grind the catalyst and disturb its even distribution.

Finally, the top support material collects scale, fines, and polymers which would otherwise plug the more densely packed catalyst. Trash baskets are installed at the top of the reactor, are surrounded by support balls, and are quite often filled with support balls.

Catalyst supports are manufactured as either balls (spheres) or cylindrical pellets. The range of properties listed in Table 8–6 are for illustrative purposes only.

SUMMARY

Hydrotreaters (HDS units) are essential to prepare feedstock for catalytic reformers. HDS units reduce naphtha contaminants such as

TABLE 8-6
Properties of catalyst supports[a]

Spheres' Diameter, in.	Ceramic Pellets Outer Diameter (OD) × Length, in.	Average Crush Strength, lb	
⅛	⅛ × ⅛	⅛-in. Spheres	50–100
¼	¼ × ⅜	½-in. Spheres	400–600
⅜	⅜ × ½	1-in. Spheres	900–1,500
½	½ × ⅝	Ceramic Pellets	13,000–15,000
⅝	⅝ × ¾		
¾	¾ × ⅞		
1	⅞ × 1¼		
1¼	–		
2	–		

Bulk Density of Catalyst Support, lb/cu ft 80–135
Chemical Composition, wt%
Al_2O_3 50–99
SiO_2 0.1–55
CaO   0.1–5

[a] Range of properties based on information available in manufacturers' bulletins

sulfur, nitrogen, oxygen, arsenic, and lead to concentrations acceptable for modern reforming catalysts.

The cobalt-moly and nickel-moly catalysts used in HDS reactors in the presence of hydrogen desulfurizes, denitrogenates, and demetallates raw-naphtha feed at temperatures of 525–850°F and at pressures from 400 to 800 psig. HDS catalysts can be regenerated 2–5 times and usually achieve ultimate lives of 300–700 bbl/lb of catalyst.

HDS units are subject to chloride and sulfide corrosion. They are also subject to fouling of exchangers, heater tubes, and the catalyst.

REFERENCES

1. Gates, Bruce C., James R. Katzer, and G.C.A. Schuit. *Chemistry of Catalytic Processes*. McGraw-Hill Book Company (1979): 406.
2. Henke, Alfred M., et al. "Hydrodesulfurization of Naphthas." U.S. pat. 3,487,011. 30 December 1969.
3. NPRA 1975 Q&A, p. 49, question 11. NPRA 1978 Q&A, p. 111, question 28 and p. 122, question 47.
4. NPRA 1978 Q&A, p. 109, question 26.
5. NPRA 1974 Q&A, p. 61, question 9. NPRA 1975 Q&A, p. 126 and NPRA 1979 Q&A, p. 85, question 39.
6. NPRA 1978 Q&A, p. 109, op. cit.
7. Johnson, M.M., and G.P. Nowack. "To Remove TEL from Hydrocarbon." *Hydrocarbon Processing* (October 1975): 119–122.

8. NPRA 1982 Q&A, p. 130, question 30.
9. Ibid.
10. NPRA 1979 Q&A, p. 107, question 29.
11. NPRA 1975 Q&A, p. 42, question 1.
12. NPRA 1979 Q&A, p. 107, question 29, op. cit.
13. NPRA 1975 Q&A, p. 42, question 1, op. cit.
14. NPRA 1977 Q&A, p. 18, question 2; p. 121, question 2; p. 124, question 2. NPRA 1978 Q&A, p. 122, question 47. NPRA 1979 Q&A, p. 108, question 32. NPRA 1982 Q&A, p. 132, question 32.
15. 1982 Ketjen Catalyst Symposium, p. 19, question B–5.
16. NPRA 1974 Q&A, p. 48, question 2.
17. NPRA 1981 Q&A, p. 121, question 17.
18. NPRA 1975 Q&A, p. 43, question 2. NPRA 1976 Q&A, p. 47, question 1; p. 48, question 2; p. 49, question 1; p. 50, question 3. NPRA 1979 Q&A, p. 102, question 22; p. 103, question 23.
19. Ibid.
20. Ibid.
21. NPRA 1975 Q&A, p. 20, question 1; p. 41, question 2. NPRA 1976 Q&A, p. 76, question 4; p. 56, question 4. NPRA 1977 Q&A, p. 84, question 10. NPRA 1978 Q&A, p. 124, question 51. NPRA 1979 Q&A, p. 104, question 25.
22. Lin, Y.R., and B.L. Crynes. "Hydroprocessing Conditions Influence Unit Fouling." *OGJ* (28 November 1983): 111.
23. Brooks, Kenneth. "How to Fight Pretreater Corrosion." *OGJ* (17 October 1960): 127.
24. Miller, R.M. "Control Initial Aqueous Condensate Corrosion." *Hydrocarbon Processing* (June 1978).
25. Brief, Richard S., Fred S. Venable, and Robert S. Ajemian. "Nickel Carbonyl: Its Detection and Potential Formation." *Industrial Hygiene Journal* (January–February 1965): 72.
26. Sunderman, William F., Jr. "Studies Firm Up Some Metals' Role In Cancer." *Chemical & Engineering News* (17 January 1977): 35.

APPENDIX I

Octane numbers of selected hydrocarbons

Compound	Formula	ASTM Octane Numbers				
		Research ml TEL/gal		Motor ml TEL/gal		
		0	3.0	0	3.0	
Paraffins						
n-Butane	C_4H_{10}	94.0	104.1	89.1	104.7	
2-Methyl Propane		102.1	118.3	97.0	—	
n-Pentane	C_5H_{12}	61.8	84.8	63.2	84.8	
2-Methylbutane		93.0	104.9	89.7	107.3	
2,2-Dimethyl Propane		85.5	101.3	80.2	99.9	
n-Hexane	C_6H_{14}	24.8	65.3	26.0	65.2	
2-Methyl Pentane		73.4	93.1	73.5	91.1	
3-Methyl Pentane		74.5	93.4	73.3	91.3	
2,2-Dimethyl Butane		91.8	106.0	93.4	113.1	
2,3-Dimethyl Butane		104.3	118.2	94.2	111.0	
n-Heptane	C_7H_{16}	0.0	41.9	0.0	48.1	
2-Methyl Hexane		42.4	73.2	46.4	74.5	
3-Methyl Hexane		52.0	74.7	55.0	81.0	
2,2-Dimethyl Pentane		92.8	104.4	95.6	114.1	
2,4-Dimethyl Pentane		83.1	96.1	83.8	99.1	
2,2,3-Trimethyl Butane		112.1	—	101.3	115.7	
n-Octane	C_8H_{18}	—	24.8	—	28.1	
2-Methyl Heptane		21.7	57.6	23.0	61.4	
4-Methyl Heptane		26.7	61.1	39.0	70.1	
3-Ethyl Hexane		33.5	61.1	52.4	80.0	
2,2-Dimethyl Hexane		72.5	93.3	77.4	95.2	
2,4-Dimethyl Hexane		65.2	87.3	69.9	89.0	

Continued.

APPENDIX I

Octane numbers of selected hydrocarbons—cont'd

Compound	Formula	ASTM Octane Numbers			
		Research ml TEL/gal		Motor ml TEL/gal	
		0	3.0	0	3.0
2,5-Dimethyl Hexane		55.5	81.6	55.7	82.9
3,4-Dimethyl Hexane		76.3	94.7	81.7	97.1
3-Methyl-3-ethyl Pentane		80.8	95.9	88.7	102.5
2,2,4-Trimethyl Pentane		100.0	115.5	100.0	115.5
2,3,3-Trimethyl Pentane		106.1	—	99.4	112.5
2,3,4-Trimethyl Pentane		102.7	—	95.9	106.7
Naphthenes					
Cyclopentane	C_5H_{10}	101.6	111.2	84.9	95.2
Methylcyclopentane	C_6H_{12}	89.3	100.9	81.0	91.5
Ethylcyclopentane	C_7H_{14}	67.2	79.5	61.2	80.7
1,1-Dimethylcyclopentane		92.3	108.0	89.3	101.3
Isopropylcyclopentane	C_8H_{16}	81.1	94.3	76.2	89.4
1,1,3-Trimethylcyclopentane		87.7	101.3	83.5	95.6
n-Butylcyclopentane	C_9H_{18}	—	29.6	—	36.7
Isobutylcyclopentane		33.4	59.2	28.2	58.1
Cyclohexane	C_6H_{12}	84.0	96.6	77.6	87.4
Methylcyclohexane	C_7H_{14}	73.8	88.7	73.8	87.4
Ethylcyclohexane	C_8H_{16}	46.5	65.1	40.8	65.4
1,1-Dimethylcyclohexane		87.3	98.0	85.9	95.7

Compound	Formula				
n-Propylcyclohexane	C₉H₁₈	17.8	42.8	14.0	47.7
Isopropylcyclohexane		62.8	79.6	61.1	81.4
1-Methyl-1-ethylcyclohexane		68.7	85.7	76.7	88.6
1,1,2-Trimethylcyclohexane		95.7	104.4	87.7	97.0
1,1,3-Trimethylcyclohexane		81.3	94.8	82.6	95.8
Isobutylcyclohexane	C₁₀H₂₀	33.7	56.4	28.9	58.3
sec-Butylcyclohexane		51.0	71.2	55.2	74.6
tert-Butylcyclohexane		98.5	107.4	89.2	96.3
Cycloheptane	C₇H₁₄	38.9	59.8	40.8	65.3
Ethylcycloheptane	C₉H₁₈	28.0	50.0	30.0	60.1
Cyclooctane	C₈H₁₆	71.0	—	58.2	—
Aromatics					
Benzene	C₆H₆	—	109.0	—	106.2
Toluene	C₇H₈	119.7	120.3	109.1	113.3
Ethylbenzene	C₈H₁₀	111.2	116.5	97.9	102.5
Orthoxylene		—	105.1	100.0	100.0
Metaxylene		117.5	120.3	115.0	120.3
Paraxylene		116.4	—	109.6	119.2
n-Propylbenzene	C₉H₁₂	111.0	118.0	98.7	—
Isopropylbenzene		113.1	—	99.3	105.3
1-Methyl-3-ethylbenzene		112.1	—	100.0	—
1,2,3-Trimethylbenzene		105.3	105.3	100.8	100.7
1,2,4-Trimethylbenzene		110.5	111.0	105.8	106.0
n-Butylbenzene	C₁₀H₁₄	104.4	111.0	94.1	101.3
Iso-Butylbenzene		111.4	115.5	98.0	103.5
1-Methyl-3-n-propylbenzene		111.9	117.2	100.5	106.0
1,2-Dimethyl-3-ethylbenzene		104.4	104.4	91.9	93.3
1,2,3,4-Tetramethylbenzene		105.3	105.3	100.3	99.7
Styrene	C₈H₈	115.5	114.3	102.5	101.3

APPENDIX II

Physical properties of important petrochemical materials

Type of Compound	Complete Name	Hydrogen, %	Carbon, %
Paraffin	Methane	25.1	74.9
Paraffin	Ethane	20.0	80.0
Paraffin	Propane	18.3	81.7
Paraffin	Normal Butane	17.3	82.7
Paraffin	Isobutane	17.3	82.7
Paraffin	Normal Pentane	16.8	83.2
Paraffin	Isopentane or 2-Methylbutane	16.8	83.2
Paraffin	Neohexane or 2,2-Dimethylbutane	16.4	83.6
Paraffin	2,3-Dimethylbutane	16.4	83.6
Paraffin	Normal Heptane	16.1	83.9
Paraffin	2,4-Dimethylpentane	16.1	83.9
Paraffin	Triptane or 2,2,3-Trimethylbutane	16.1	83.9
Paraffin	Isooctane (Octane) or 2,2,4-Trimethylpentane	15.9	84.1
Paraffin	2,2,3-Trimethylpentane	15.9	84.1
Paraffin	2,3,3-Trimethylpentane	15.9	84.1
Paraffin	2,3,4-Trimethylpentane	15.9	84.1
Paraffin	Normal Decane	15.6	84.4
Paraffin	Cetane	15.1	84.9
Cyclic Paraffin	Cyclopentane	14.4	85.6
Cyclic Paraffin	Methylcyclopentane	14.4	85.6
Cyclic Paraffin	Cyclohexane	14.4	85.6
Cyclic Paraffin or Naphthene	Decalin	13.1	86.9
Cyclic Paraffin or Naphthene	2-Cyclohexyl-octane	14.4	85.6
Aromatic	Benzene	7.7	92.3
Aromatic	Toluene	8.8	91.2
Aromatic	Ortho-xylene or 1,2-Dimethylbenzene	9.5	90.5
Aromatic	Meta-xylene or 1,3-Dimethylbenzene	9.5	90.5
Aromatic	Para-xylene or 1,4-Dimethylbenzene	9.5	90.5
Aromatic	Ethylbenzene	9.5	90.5
Aromatic	Cumene or Isopropylbenzene	10.1	89.9
Aromatic	Normal Propyl Benzene	10.1	89.9
Aromatic	Naphthalene (Moth Balls)	6.3	93.7
Aromatic	Alpha-methyl-naphthalene	7.1	92.9
Aromatic	2-Phenyl-octane	11.6	88.4
Aromatic Amine	Aniline	7.6	77.4

(Source: O&GJ, 20 October 1952, p. 173)

APPENDIX 199

| Boiling Point, °F | Rvp, psi | Latent Heat of Evaporation, BTU/lb | Freezing Point, °F | Heating Value, BTU/lb | At 68° F |||
					BTU/U.S. gal	Specific Gravity	Wt/U.S. gal, lb
−259	—	219	−296	21,500	—	Gas	Gas
−128	—	210	−298	20,400	—	Gas	Gas
−44	—	183	−306	19,800	—	Gas	Gas
31	—	165	−217	19,500	94,000	0.580	4.83
11	—	158	−255	19,500	91,000	0.558	4.65
97	16	154	−201	19,300	101,000	0.627	5.23
82	22	146	−256	19,300	100,000	0.621	5.17
122	10	131	−148	19,200	104,000	0.650	5.42
136	8	136	−199	19,200	106,000	0.663	5.52
209	1.7	136	−131	19,200	110,000	0.685	5.71
177	3.5	127	−183	19,100	107,000	0.674	5.61
178	3.5	124	−13	19,100	110,000	0.691	5.76
211	1.8	117	−161	19,100	110,000	0.693	5.77
230	1.2	121	−170	19,100	114,000	0.717	5.98
239	0.9	123	−149	19,100	116,000	0.727	6.06
236	1.0	123	−165	19,100	115,000	0.720	6.00
345	0.06	119	−21	19,000	116,000	0.731	6.09
547	<0.001	98	+65	18,900	122,000	0.774	6.45
121	10.5	167	−137	18,800	117,000	0.746	6.22
161	4.8	148	−224	18,800	118,000	0.750	6.25
177	3.3	154	+44	18,700	122,000	0.780	6.50
374	0.04	135	−40	20,000	140,000	0.880	7.40
504	<0.001	135	<−70	—	—	0.825	6.88
176	3.2	169	+42	17,300	127,000	0.881	7.34
231	1.2	156	−139	17,400	126,000	0.869	7.23
292	0.3	149	−13	17,500	129,000	0.882	7.35
282	0.35	147	−54	17,500	126,000	0.866	7.21
281	0.35	146	+56	17,500	126,000	0.863	7.19
277	0.4	146	−139	17,600	127,000	0.869	7.24
306	0.2	134	−141	17,700	127,000	0.864	7.19
319	0.15	137	−147	17,700	127,000	0.864	7.19
424	Solid	—	+176	16,700	—	Solid	Solid
473	0.003	140	−23	16,700	140,000	1.022	8.51
464	0.001	139	−38	18,200	130,000	0.859	7.15
364	0.04	187	+21	15,000	128,000	1.024	8.53

Continued.

APPENDIX II

Physical properties of important petrochemical materials—cont'd

Type of Compound	Complete Name	Hydrogen, %	Carbon, %
Aromatic Amine	Monomethyl Aniline or n-Methylaniline	8.5	78.5
Aromatic Amine	2,4-Xylidine	9.2	79.3
Aromatic Amine	2,6-Xylidine	9.2	79.3
Metal-organic	Tetraethyl lead	6.2	29.7
Organic Halide	Ethylene Dibromide (1,2-Dibromoethane)	2.1	12.8
Alcohol	Ethanol (Grain Alcohol)	13.1	52.1
Alcohol	Isopropanol (Isopropyl Alcohol)	13.4	60.0
Aromatic-olefin	Styrene	7.7	92.3
Olefin	Ethylene	14.4	85.6
Olefin	Butylene-1, 1-butene	14.4	85.6
Olefin	Butylene-2, cis-2-butene	14.4	85.6
Olefin	1-Pentene	14.4	85.6
Olefin	1-Heptene	14.4	85.6
Olefin	1-Decene	14.4	85.6
Olefin	1-Dodecene	14.4	85.6
Olefin	Cetene or 1-Hexadecene	14.4	85.6
Olefin	Iso-butylene or 2-Methylpropene	14.4	85.6
Olefin	2,3,3-Trimethyl-1-butene (Triptene)	14.4	85.6
Olefin	Diisobutylene or 2,4,4-Trimethyl-1-pentene	14.4	85.6
Olefin	Diisobutylene or 2,4,4-Trimethyl-2-pentene	14.4	85.6
Diolefin	1,3-Butadiene	11.2	88.8
Acetylene	Acetylene	7.7	92.3
Acetylene	1-Butyne	11.2	88.8
Organic Nitrate	Amyl Nitrate	9.5	51.3
Organic Sulfur Compound	Butyl Mercaptan	11.2	53.3
	Sulfur	—	—
	Hydrogen Sulfide	5.9	—

(Source: O&GJ, 20 October 1952, p. 173)

APPENDIX

Boiling Point, °F	Rvp, psi	Latent Heat of Evaporation, BTU/lb	Freezing Point, °F	Heating Value, BTU/lb	At 68° F		
					BTU/U.S. gal	Specific Gravity	Wt/U.S. gal, lb
384	0.02	172	−71	15,600	128,000	0.988	8.23
420	0.005	150	—	15,600	127,000	0.976	8.13
422	0.005	150	—	15,600	127,000	0.981	8.17
388	0.02	60	−202	7,800	108,000	1.656	13.80
269	0.5	83	+50	—	—	2.185	18.2
173	2.3	370	−174	11,500	76,000	0.790	6.58
180	1.8	290	−129	13,100	86,000	0.786	6.55
293	0.3	152	−23	17,400	132,000	0.908	7.56
−155	—	208	−272	20,300	—	Gas	Gas
+21	62	168	−302	19,500	97,000	0.596	4.97
+39	46	179	−218	19,400	101,000	0.622	5.19
+86	19	154	−265	19,300	104,000	0.642	5.35
201	2.0	148	−182	19,200	112,000	0.698	5.82
339	0.09	131	−98	19,100	118,000	0.742	6.18
416	0.015	123	−32	19,100	121,000	0.759	6.33
545	<0.001	112	+39	19,000	124,000	0.783	6.52
20	65	169	−221	19,400	96,000	0.595	4.96
172	3.6	124	−169	19,100	112,000	0.706	5.88
215	1.6	—	−136	19,000	113,000	0.716	5.97
221	1.4	—	−160	19,000	114,000	0.722	6.02
24	59	178	−164	19,000	98,000	0.622	5.18
−119	—	—	−114	20,700	—	Gas	Gas
+48	38	186	−194	19,600	106,000	0.65	5.4
305	0.18	121	−139	—	—	0.999	8.32
208	0.9	151	−177	15,000	105,000	0.836	6.97
832	—	—	+235	3,980	—	Solid	Solid
−77	—	—	−122	6,500	—	Gas	Gas

APPENDIX III

Wire mesh openings (U.S. sieve series)

Sieve Number	Sieve Opening	
	in.	mm
2.5	0.3150	8.00
3.0	0.2650	6.73
3.5	0.2230	5.66
4.0	0.1870	4.75
5.0	0.1570	3.99
6.0	0.1320	3.35
7.0	0.1110	2.82
8.0	0.0937	2.38
10.0	0.0787	2.00
14.0	0.0555	1.41
18.0	0.0394	1.00
20.0	0.0331	0.84
25.0	0.0280	0.71
30.0	0.0232	0.59
40.0	0.0165	0.42

(Source: *Chemical Engineers Handbook*, 2d ed., edited by John H. Perry, © McGraw-Hill, 1941, p. 1720)

APPENDIX IV

Temperature conversion chart

How to Use These Tables

Celsius to Fahrenheit: Use the center column as Celsius temperature and read the Fahrenheit temperature in the °F column adjacent and to the right

Fahrenheit to Celsius: Read the center or base column as °F and use the corresponding °C column to the left for your converted temperature

Conversion Formula: Celsius to Fahrenheit (°C × 9/5) + 32 = °F
Fahrenheit to Celsius (°F − 32) × 5/9 = °C

To °C	Base °F or °C	To °F	To °C	Base °F or °C	To °F
−73.33	−100	−148.0	−21.11	−6	21.2
−70.55	−95	−139.0	−20.56	−5	23.0
−67.78	−90	−130.0	−20.00	−4	24.8
−65.00	−85	−121.0	−19.44	−3	26.6
−62.22	−80	−112.0	−18.89	−2	28.4
−59.45	−75	−103.0	−18.33	−1	30.2
−56.67	−70	−94.0	−17.78	0	32.0
−53.89	−65	−85.0	−17.22	1	33.8
−51.11	−60	−76.0	−16.67	2	35.6
−48.34	−55	−67.0	−16.11	3	37.4
−45.56	−50	−58.0	−15.56	4	39.2
−42.77	−45	−49.0	−15.00	5	41.0
−40.00	−40	−40.0	−14.44	6	42.8
−37.23	−35	−31.0	−13.89	7	44.6
−34.44	−30	−22.0	−13.33	8	46.4
−31.66	−25	−13.0	−12.78	9	48.2
−31.11	−24	−11.2	−12.22	10	50.0
−30.55	−23	−9.4	−11.67	11	51.8
−30.00	−22	−7.6	−11.11	12	53.6
−29.45	−21	−5.8	−10.56	13	55.4
−28.89	−20	−4.0	−10.00	14	57.2
−28.34	−19	−2.2	−9.44	15	59.0
−27.78	−18	−0.4	−8.89	16	60.8
−27.23	−17	1.4	−8.33	17	62.6
−26.67	−16	3.2	−7.78	18	64.4
−26.12	−15	5.0	−7.22	19	66.2
−25.56	−14	6.8	−6.67	20	68.0
−25.00	−13	8.6	−6.11	21	69.8
−24.44	−12	10.4	−5.56	22	71.6
−23.88	−11	12.2	−5.00	23	73.4
−23.33	−10	14.0	−4.44	24	75.2
−22.77	−9	15.8	−3.89	25	77.0
−22.22	−8	17.6	−3.33	26	78.8
−21.67	−7	19.4	−2.78	27	80.6

(courtesy Kaiser Chemical)

Continued.

APPENDIX

To °C	Base °F or °C	To °F	To °C	Base °F or °C	To °F
−2.22	28	82.4	26.67	80	176.0
−1.67	29	84.2	27.22	81	177.8
−1.11	30	86.0	27.78	82	179.6
−0.56	31	87.8	28.33	83	181.4
0.00	32	89.6	28.89	84	183.2
0.56	33	91.4	29.44	85	185.0
1.11	34	93.2	30.00	86	186.8
1.67	35	95.0	30.56	87	188.6
2.22	36	96.8	31.11	88	190.4
2.78	37	98.6	31.67	89	192.2
3.33	38	100.4	32.22	90	194.0
3.89	39	102.2	32.78	91	195.8
4.44	40	104.0	33.33	92	197.6
5.00	41	105.8	33.89	93	199.4
5.56	42	107.6	34.44	94	201.2
6.11	43	109.4	35.00	95	203.0
6.67	44	111.2	35.56	96	204.8
7.22	45	113.0	36.11	97	206.6
7.78	46	114.8	36.67	98	208.4
8.33	47	116.6	37.22	99	210.2
8.89	48	118.4	37.78	100	212.0
9.44	49	120.2	38.33	101	213.8
10.00	50	122.0	38.89	102	215.6
10.56	51	123.8	39.44	103	217.4
11.11	52	125.6	40.00	104	219.2
11.67	53	127.4	40.56	105	221.0
12.22	54	129.2	41.11	106	222.8
12.78	55	131.0	41.67	107	224.6
13.33	56	132.8	42.22	108	226.4
13.89	57	134.6	42.78	109	228.2
14.44	58	136.4	43.33	110	230.0
15.00	59	138.2	43.89	111	231.8
15.56	60	140.0	44.44	112	233.6
16.11	61	141.8	45.00	113	235.4
16.67	62	143.6	45.56	114	237.2
17.22	63	145.4	46.11	115	239.0
17.78	64	147.2	46.67	116	240.8
18.33	65	149.0	47.22	117	242.6
18.89	66	150.8	47.78	118	244.4
19.44	67	152.6	48.33	119	246.2
20.00	68	154.4	48.89	120	248.0
20.56	69	156.2	49.44	121	249.8
21.11	70	158.0	50.00	122	251.6
21.67	71	159.8	50.56	123	253.4
22.22	72	161.6	51.11	124	255.2
22.78	73	163.4	51.67	125	257.0
23.33	74	165.2	52.22	126	258.8
23.89	75	167.0	52.78	127	260.6
24.44	76	168.8	53.33	128	262.4
25.00	77	170.6	53.89	129	264.2
25.56	78	172.4	54.44	130	266.0
26.11	79	174.2	55.00	131	267.8

APPENDIX 205

To °C	Base °F or °C	To °F	To °C	Base °F or °C	To °F
55.56	132	269.6	84.44	184	363.2
56.11	133	271.4	85.00	185	365.0
56.67	134	273.2	85.56	186	366.8
57.22	135	275.0	86.11	187	368.6
57.78	136	276.8	86.67	188	370.4
58.33	137	278.6	87.22	189	372.2
58.89	138	280.4	87.78	190	374.0
59.44	139	282.2	88.33	191	375.8
60.00	140	284.0	88.89	192	377.6
60.56	141	285.8	89.44	193	379.4
61.11	142	287.6	90.00	194	381.2
61.67	143	289.4	90.56	195	383.0
62.22	144	291.2	91.11	196	384.8
62.78	145	293.0	91.67	197	386.6
63.33	146	294.8	92.22	198	388.4
63.89	147	296.6	92.78	199	390.2
64.44	148	298.4	93.33	200	392.0
65.00	149	300.2	93.89	201	393.8
65.56	150	302.0	94.44	202	395.6
66.11	151	303.8	95.00	203	397.4
66.67	152	305.6	95.56	204	399.2
67.22	153	307.4	96.11	205	401.0
67.78	154	309.2	96.67	206	402.8
68.33	155	311.0	97.22	207	404.6
68.89	156	312.8	97.78	208	406.4
69.44	157	314.6	98.33	209	408.2
70.00	158	316.4	98.89	210	410.0
70.56	159	318.2	99.44	211	411.8
71.11	160	320.0	100.00	212	413.6
71.67	161	321.8	100.56	213	415.4
72.22	162	323.6	101.11	214	417.2
72.78	163	325.4	101.67	215	419.0
73.33	164	327.2	102.22	216	420.8
73.89	165	329.0	102.78	217	422.6
74.44	166	330.8	103.33	218	424.4
75.00	167	332.6	103.89	219	426.2
75.56	168	334.4	104.44	220	428.0
76.11	169	336.2	107.22	225	437.0
76.67	170	338.0	110.00	230	446.0
77.22	171	339.8	112.78	235	455.0
77.78	172	341.6	115.56	240	464.0
78.33	173	343.4	118.33	245	473.0
78.89	174	345.2	121.11	250	482.0
79.44	175	347.0	123.89	255	491.0
80.00	176	348.8	126.67	260	500.0
80.56	177	350.6	129.44	265	509.0
81.11	178	352.4	132.22	270	518.0
81.67	179	354.2	135.00	275	527.0
82.22	180	356.0	137.78	280	536.0
82.78	181	357.8	140.56	285	545.0
83.33	182	359.6	143.33	290	554.0
83.89	183	361.4	146.11	295	563.0

Continued.

To °C	Base °F or °C	To °F	To °C	Base °F or °C	To °F
148.89	300	572.0	382.22	720	1328.0
151.67	305	581.0	387.78	730	1346.0
154.44	310	590.0	393.33	740	1364.0
157.22	315	599.0	398.89	750	1382.0
160.00	320	608.0	404.44	760	1400.0
162.78	325	617.0	410.00	770	1418.0
165.56	330	626.0	415.56	780	1436.0
168.33	335	635.0	421.11	790	1454.0
171.11	340	644.0	426.27	800	1472.0
173.89	345	653.0	432.22	810	1490.0
176.67	350	662.0	437.78	820	1508.0
179.44	355	671.0	443.33	830	1526.0
182.22	360	680.0	448.89	840	1544.0
185.00	365	689.0	454.44	850	1562.0
187.78	370	698.0	460.00	860	1580.0
190.56	375	707.0	465.56	870	1598.0
193.33	380	716.0	471.11	880	1616.0
196.11	385	725.0	476.67	890	1634.0
198.89	390	734.0	482.22	900	1652.0
201.67	395	743.0	487.78	910	1670.0
204.44	400	752.0	493.33	920	1688.0
210.00	410	770.0	498.89	930	1706.0
215.56	420	788.0	504.44	940	1724.0
221.11	430	806.0	510.00	950	1742.0
226.67	440	824.0	515.56	960	1760.0
232.22	450	842.0	521.11	970	1778.0
237.78	460	860.0	526.67	980	1796.0
243.33	470	878.0	532.22	990	1814.0
248.89	480	896.0	537.78	1000	1832.0
254.44	490	914.0	543.33	1010	1850.0
260.00	500	932.0	548.89	1020	1868.0
265.56	510	950.0	554.44	1030	1886.0
271.11	520	968.0	560.00	1040	1904.0
276.67	530	986.0	565.56	1050	1922.0
282.22	540	1004.0	571.11	1060	1940.0
287.78	550	1022.0	576.67	1070	1958.0
293.33	560	1040.0	582.22	1080	1976.0
298.89	570	1058.0	587.78	1090	1994.0
304.44	580	1076.0	593.33	1100	2012.0
310.00	590	1094.0	598.89	1110	2030.0
315.56	600	1112.0	604.44	1120	2048.0
321.11	610	1130.0	610.00	1130	2066.0
326.67	620	1148.0	615.56	1140	2084.0
332.22	630	1166.0	621.11	1150	2102.0
337.78	640	1184.0	626.67	1160	2120.0
343.33	650	1202.0	632.22	1170	2138.0
348.89	660	1220.0	637.78	1180	2156.0
354.44	670	1238.0	643.33	1190	2174.0
360.00	680	1256.0	648.89	1200	2192.0
365.56	690	1274.0	654.44	1210	2210.0
371.11	700	1292.0	660.00	1220	2228.0
376.67	710	1310.0	665.56	1230	2246.0

To °C	Base °F or °C	To °F	To °C	Base °F or °C	To °F
671.11	1240	2264.0	887.78	1630	2966.0
676.67	1250	2282.0	893.33	1640	2984.0
682.22	1260	2300.0	898.89	1650	3002.0
687.78	1270	2318.0	904.44	1660	3020.0
693.33	1280	2336.0	910.00	1670	3038.0
698.89	1290	2354.0	915.56	1680	3056.0
704.44	1300	2372.0	921.11	1690	3074.0
710.00	1310	2390.0	926.67	1700	3092.0
715.56	1320	2408.0	932.22	1710	3110.0
721.11	1330	2426.0	937.78	1720	3128.0
726.67	1340	2444.0	943.33	1730	3146.0
732.22	1350	2462.0	948.89	1740	3164.0
737.78	1360	2480.0	954.44	1750	3182.0
743.33	1370	2498.0	960.00	1760	3200.0
748.89	1380	2516.0	965.56	1770	3218.0
754.44	1390	2534.0	971.11	1780	3236.0
760.00	1400	2552.0	976.67	1790	3254.0
765.56	1410	2570.0	982.22	1800	3272.0
771.11	1420	2588.0	987.78	1810	3290.0
776.67	1430	2606.0	993.33	1820	3308.0
782.22	1440	2624.0	998.89	1830	3326.0
787.78	1450	2642.0	1004.00	1840	3344.0
793.33	1460	2660.0	1010.00	1850	3362.0
798.89	1470	2678.0	1016.00	1860	3380.0
804.44	1480	2696.0	1021.00	1870	3398.0
810.00	1490	2714.0	1027.00	1880	3416.0
815.56	1500	2732.0	1032.00	1890	3434.0
821.11	1510	2750.0	1038.00	1900	3452.0
826.67	1520	2768.0	1043.00	1910	3470.0
832.22	1530	2786.0	1049.00	1920	3488.0
837.78	1540	2804.0	1054.00	1930	3506.0
843.33	1550	2822.0	1060.00	1940	3524.0
848.89	1560	2840.0	1066.00	1950	3542.0
854.44	1570	2858.0	1071.00	1960	3560.0
860.00	1580	2876.0	1077.00	1970	3578.0
865.56	1590	2894.0	1082.00	1980	3596.0
871.11	1600	2912.0	1088.00	1990	3614.0
876.67	1610	2930.0	1093.00	2000	3632.0
882.22	1620	2948.0			

APPENDIX V

1984 Licensors of commercial catalytic reforming processes

- IFP Catalytic Reforming
  Institut Francais du Petrole
  680 Fifth Avenue
  New York, NY 10019

- Magnaforming
  Engelhard Industries Div.
  429 Delancey St.
  Newark, NJ 07105

- Platforming
  UOP
  UOP Process Div.
  20 UOP Plaza
  Algonquin & Mt. Prospect Roads
  Des Plaines, IL 60016

- Powerforming
  Exxon Research and Engineering Co.
  Suite 615
  1200 Smith St.
  Houston, TX 77002–4590

- Rheniforming
  Chevron Research Co.
  525 Market St.
  San Francisco, CA 94105

- Ultraforming
  Standard Oil Co. (Indiana)
  P.O. Box 5910A
  Chicago, IL 60680

APPENDIX VI

Manufacturers of reforming catalysts

American Cyanamid Co.
Refinery Catalysts
One Cyanamid Plaza
Wayne, NJ 07470

Chevron Research Co.
525 Market St.
San Francisco, CA 94105

Cyanamid Ketjen
Akzo-Chemie Nederland, B.V.
Station Straat 48
3800 Az Amersfoort
The Netherlands

Exxon Research and Engineering
Suite 615
1200 Smith St.
Houston, TX 77002–4590

Engelhard Corp.
429 Delancey St.
Newark, NJ 07105

Instituto Mexican del Petroleo
Av. de los Cien Metros No. 152
Directo 587–13–71
Mexico 14, D.F.
Mexico

Katalysatorewerke Huels Ag
D–4370 Marl
West Germany

Procatalyse
IFP Enterprises Inc.
450 Park Ave.
New York, NY 10022
      or
Institut Francais du Petrole
1 et 4, Avenue de Bois-Preau
92506 Rueil-Malmaison
France

UOP Process Division
20 UOP Plaza
Algonquin & Mt. Prospect Roads
Des Plaines, IL 60016

APPENDIX VII

Hydrocarbons in straight-run base stocks (97°–243°F boiling range), vol %

State, Field, and Crude Oil Source	Cyclo-pentane	Methyl-cyclo-pentane	Benzene	Cyclo-hexane	1,1-Di-methyl-cyclo-pentane
California					
Coalinga, Eastside (Eocene)..	0.13	0.76	0.17	0.56	0.09
Louisiana					
Golden Meadow B........	0.04	0.23	0.05	0.27	0.03
Jennings (Evangeline)......	0.04	0.29	0.21	0.41	0.06
Texas					
Carthage (Upper Pettit)[a]	0.10	0.73	2.08	1.18	0.07
Chapel Hill[a]............	0.10	0.77	0.38	0.61	0.08
Conroe................	0.10	0.71	0.36	1.13	0.04
East Texas..............	0.12	0.75	0.07	0.49	0.04
Hastings...............	0.05	0.31	0.01	0.56	0.03
Hulk-Silk-Sikes..........	0.17	0.87	0.06	0.47	0.12
KMA (Strawn)	0.18	1.09	0.09	0.68	0.06
Old Ocean..............	0.10	0.65	0.27	0.88	0.10
Plymouth	0.10	0.62	0.05	1.35	0.07
Saxet.................	0.01	0.11	None	0.22	0.02
Segno	0.10	0.63	0.18	0.95	0.05
Tom O'Connor..........	0.11	0.72	0.06	1.18	0.07
Wade City.............	0.10	0.56	0.03	1.13	0.07
Wasson................	0.14	0.76	0.80	0.89	0.07
Yates, Taylor Link	0.05	0.24	None	0.27	0.07
Wyoming					
Elk Basin	0.09	0.36	0.08	0.27	0.07

[a]Condensate

(Source: OGJ, 20 October 1952, p. 174, Table 2)

trans-1,3-Di-methyl-cyclo-pentane	trans-1,2-Di-methyl-cyclo-pentane	Methyl-cyclo-hexane	Ethyl-cyclo-pentane	Toluene	Tri-methyl-cyclo-pentanes	Alkyl cyclo-pentanes	Total
0.36	0.52	1.07	0.32	0.60	0.53	0.07	7.39
0.20	0.14	0.63	0.15	0.17	0.20	0.06	4.09
0.15	0.05	1.05	0.14	0.72	0.22	0.03	5.82
0.36	0.27	1.97	0.42	0.87	0.48	None	29.30
0.22	0.13	1.48	0.29	0.37	0.28	0.05	24.18
0.28	0.17	2.40	0.22	1.78	0.39	0.01	10.89
0.53	0.51	1.17	0.28	0.30	0.39	0.13	8.87
0.24	0.10	1.33	0.15	0.03	0.13	0.04	4.12
0.66	1.20	1.21	0.57	0.29	0.70	0.19	13.94
1.03	0.77	1.79	0.53	0.37	0.66	0.16	14.52
0.47	0.17	2.07	0.42	0.72	0.41	0.05	12.03
0.48	0.34	2.77	0.42	0.25	0.37	0.14	10.60
0.07	0.05	0.59	0.06	None	0.07	0.04	1.58
0.30	0.12	2.14	0.27	0.98	0.25	0.03	9.90
0.47	0.28	2.48	0.35	0.36	0.39	0.13	10.76
0.30	0.30	2.69	0.39	0.23	0.28	0.12	7.56
0.44	0.36	1.16	0.32	1.52	0.52	0.05	13.22
0.56	0.13	0.59	0.06	None	0.48	0.20	6.24
0.31	0.23	0.68	0.25	0.49	0.25	0.01	9.84

APPENDIX VIII

Hydrocarbons in midcontinent naphthas

Carbon No.	Hydrocarbon	Boiling Point, °C	% Present in Naphthas		
			Virgin	Thermal	Catalytic
4	2-Methylpropane	−11.7	8.7	0.7	2.7
	2-Methylpropene	−6.9	—	1.1	1.7
	1-Butene	−6.3	—	1.5	4.9
	n-Butane	−0.5	91.3	66.6	27.0
	trans-2-Butene	0.9	—	30.1	27.1
	cis-2-Butene	3.7	—		36.6
			100.0	100.0	100.0
5	2,2-Dimethylpropane	9.5	—	0.1	0.6
	3-Methyl-1-butene	20.2	—	2.4	51.8
	2-Methylbutane	27.8	42.1	20.4	2.1
	1-Pentene	30.1	—	12.6	7.6
	2-Methyl-1-butene	31.1	—	8.2	9.2
	trans-2-Pentene	36.0	—	9.5	8.7
	n-Pentane	36.1	53.0	35.2	17.3
	2-Methyl-2-butene	38.5	—	6.8	1.8
	Cyclopentene	(44.0)	—	2.4	0.9
	Cyclopentane	49.3	4.9	2.4	100.0
			100.0	100.0	
6	2,2-Dimethylbutane	49.7	0.5	0.3	0.2
	3 and 4-Methyl-1-pentene	54.0	—	5.4	1.5
	2,3-Dimethyl-1-butene	55.6	—	—	1.5
	2,3-Dimethylbutane	58.0	0.5	1.1	9.0
	cis-4-Methyl-2-pentene	58.4	—	3.8	2.6
	2-Methylpentane	60.3	17.6	10.7	29.1
	2-Methyl-1-pentene	62.2	—	3.8	5.0

APPENDIX   213

Compound				
3-Methylpentane	63.3	10.5	7.7	18.5
1-Hexene	63.6	—	8.8	0.4
3-Methyl-1-cyclopentene	(64.9)	—	1.9	1.9
2-Ethyl-1-butene	65.0	—	0.4	2.8
2-Methyl-2-pentene	67.2	—	3.8	5.6
trans-3-Methyl-2-pentene	67.8	—	2.3	2.7
cis and trans-2 and 3-Hexene	67.6–68.6	—	6.9	4.9
n-Hexane	68.7	38.7	23.0	5.2
cis-3-Methyl-2-pentene	70.5	—	0.8	1.4
Methylcyclopentane	71.8	18.6	7.3	3.7
2,3-Dimethyl-2-butene	73.2	—	0.3	0.4
1 and 4-Methyl-1-cyclopentene	(75.8–76.0)	—	5.8	1.5
Benzene	80.1	1.0	1.3	0.9
Conjugated Diolefins	—	—	0.1	0.1
Nonconjugated Diolefins	—	—		0.1
Cyclohexane	80.7	12.6	3.4	0.7
Cyclohexene	(83.2)	—	1.1	0.3
		100.0	100.0	100.0
Olefins	76–78	—	—	0.7
2,2-Dimethylpentane	79.2	1.1	—	} 1.9
2,4-Dimethylpentane	80.5	0.4	—	
Olefins	81–86	—	—	4.0
3,3-Dimethylpentane	86.6	—	—	0.7
1,1-Dimethylcyclopentane	87.5	0.8	—	0.2
Olefins	87–89	—	—	2.7
2,3-Dimethylpentane	87.8	0.8	—	16.9
2-Methylhexane	90.1	10.1	—	6.5
trans-1,3-Dimethylcyclopentane	90.8	4.1	—	0.7
Olefins	91.0	—	—	3.5
trans-1,2-Dimethylcyclopentane	91.9	13.1	—	7.9
3-Methylhexane	92.0	13.9	—	11.5
Conjugated Diolefins	—	—	—	0.4
Nonconjugated Diolefins	—	—	—	0.2
Olefins	93–99	—	—	17.5

(Source: O&GJ, 20 October 1952, p. 175, Table 4.)

APPENDIX VIII

Hydrocarbons in midcontinent naphthas—cont'd

| Carbon No. | Hydrocarbon | Boiling Point, °C | % Present in Naphthas |||
			Virgin	Thermal	Catalytic
	n-Heptane	98.4	26.2	—	4.4
	cis-1,2-Dimethylcyclopentane	99.3	—	—	1.8
	Methylcyclohexane	100.9	22.1	—	7.1
	3 and 4-Methyl-1-cyclohexane	(102–103)	—	—	2.4
	Ethylcyclopentane	103.5	3.7	—	1.4
	1-Methyl-1-cyclohexene	(109.5)	—	—	0.7
	Toluene	110.6	3.7	—	6.9
			100.0		100.0
8	Paraffins	—	56.6	—	41.8
	Olefins and Cycloolefins	—	—	—	33.7
	1,1,3-Trimethylcyclopentane	(105)	} 6.8	—	4.2
	Trimethylcyclopentanes	—		—	2.1
	Trimethylcyclopentenes	—	—	—	6.5
	Dimethylcyclohexanes	—	14.9	—	} 7.0
	trans-1,2-Dimethylcyclohexane	123.4	5.1	—	
	Ethylcyclohexane	131.8	9.8	—	
	Aromatics	—	6.8	—	4.7
			100.0		100.0
9	Paraffins	—	52.1	—	67.0
	Olefins and Cycloolefins	—	—	—	16.3
	Cycloparaffins	—	41.0	—	9.7
	Aromatics	—	6.9	—	7.0
			100.0		100.0

APPENDIX IX

Properties of naphthas in world crude oils[1]

Country and Crude Name	Crude Gravity, °API	Approximate Boiling Range, °F	Naphthas in Crude			
			In Crude, Vol %	P/N/A, Vol %	N + 2A, Vol %	Sulfur, Wt %
Abu Dhabi						
Abu Albu Khoosh	31.6	150–330	16.8	68.3/18.5/13.2	44.9	0.05
Algeria						
Saharan Blend	45.5	160–290	18.3	70.8/21.8/7.1	36.0	—
		290–390	13.7	47.4/36.4/15.9	68.2	—
Zarzaitine	43.0	180–290	15.1	59.4/33.9/6.7	47.3	0.0138
Angola						
Cabinda	31.7	170–310	10.0	50.0/44.8/5.2	54.4	0.06
Soyo Blend	33.7	180–300	9.6	—/37.2/10.0	57.2	—
Brunei						
Champion Export	23.9	300–390	8.1	15.0/69.0/16.0	—	—
Seria Light	36.2	220–390	21.4	35.9/34.6/17.7	70.0	—
Cameroon						
Kole Marine Blend	34.9	180–360	21.9	38.5/53.3/8.2	69.7	0.0080
Canada						
Bow River Heavy	26.7	C_5–370	20.4	65.5/26.8/7.3	41.4	0.0061
Federated Light	39.7	C_5–370	30.9	28.3/54.9/16.8	88.5	0.005
Gulf Alberta	35.1	C_5–370	30.6	56.0/31.5/12.4	56.3	0.036
Lloydminster Light & Medium	20.7	C_5–370	19.5	55.1/31.6/11.8	55.2	0.079
Rainbow	40.7	C_5–370	38.7	61.3/26.0/12.8	51.6	0.014
Rangeland South	39.5	C_5–370	38.8	42.5/36.9/20.7	78.3	0.012

Properties of naphthas in world crude oils—cont'd

Country and Crude Name	Crude Gravity, °API	Approximate Boiling Range, °F	Naphthas in Crude			
			In Crude, Vol %	P/N/A, Vol %	N + 2A, Vol %	Sulfur, Wt %
China						
Shengli	24.2	160–290	3.3	45.9/44.7/9.4	63.5	—
Shengli		290–390	3.7	48.9/30.9/20.2	71.3	0.05
Taching	33.0	160–290	5.4	46.6/50.4/2.6	55.6	0.013
		290–390	5.8	39.6/52.4/6.5	65.4	0.010
Congo						
Djeno Blend	26.9	180–360	11.9	52.5/39.8/7.7	55.2	0.16
Denmark						
Gorm, North Sea	33.9	C_5–300	19.7	53.0/40.0/7.0	54.0	0.001
Gorm, North Sea		300–400	10.0	32.0/47.0/21.0	89.0	0.026
Divided Zone						
Burgan (Wafra)	23.3	200–435	13.6	44.0/36.0/20.0	76.0	0.15
Eocene (Wafra)	18.6	200–435	11.8	45.0/40.0/15.0	70.0	0.27
Hout	32.8	C_5–370	24.5	70.8/15.6/13.6	42.8	0.035
Khafji	28.5	C_5–370	20.4	73.4/15.4/11.2	37.8	0.026
Dubai, Uae						
Dubai (Fateh)	31.5	60–290	16.4	68.4/23.1/8.5	40.4	0.019
Ecuador						
Oriente	29.2	IBP–200	3.9	62.4/35.0/2.6	40.2	0.001
Oriente		200–400	17.3	44.9/45.8/9.3	64.4	0.01
Egypt						
Belayim	27.5	180–320	10.2	57.0/33.0/10.0	53.0	—
Gulf of Suez Mix	31.9	65–200	7.8	80.0/18.0/2.0	22.0	0.018
Gulf of Suez Mix		200–320	10.0	57.3/32.7/10.1	52.9	0.022
Gabon						
Gamba	31.8	180–320	4.9	46.9/50.6/2.5	55.6	0.007
Lucina Marine	39.5	180–300	13.5	54.3/37.8/7.9	53.6	0.005
Mandji	30.5	180–300	9.6	55.2/36.7/8.1	52.9	0.006
India						
Bombay High	39.4	C_5–300	18.6	53.7/25.0/21.3	67.6	3 wt ppm

Indonesia					
Ardjuna	35.2	150–380	27.2	34.3/46.0/19.7	85.4	0.012
Arimbi	31.8	230–335	11.2	42.3/47.0/10.7	68.4	0.030
Attaka	42.3	210–300	17.6	41.0/35.0/24.7	84.4	—
Bekapai	40.0	C₅–210	9.3	45.8/47.6/6.6	60.8	<1 wt ppm
Bekapai		210–365	26.5	33.2/48.9/17.9	84.7	68 wt ppm
Cinta	33.4	180–300	4.8	55.2/39.0/4.9	48.8	5 wt ppm
Cinta		300–380	4.0	67.6/22.6/8.1	38.8	12 wt ppm
Salawati	38.0	200–300	10.7	71.4/27.9/0.8	29.5	—
Walio	34.1	200–370	19.0	63.3/30.7/6.0	42.7	—
Minas	34.5	200–340	9.4	62.0/30.0/8.0	46.0	0.002
Iran						
Aboozar	26.9	C₅–200	5.5	72.6/24.2/3.1	30.4	0.090
Aboozar		200–320	8.4	60.0/27.3/12.7	40.7	0.097
Aboozar		320–360	3.8	59.1/25.8/15.1	56.0	0.158
Dorrood	33.6	C₅–200	10.1	85.6/12.2/2.2	16.6	0.067
Dorrood		200–320	12.7	67.5/17.8/14.8	47.4	0.088
Iranian Heavy	31.0	150–300	14.5	53.4/33.1/11.4	60.1	0.088
Iranian Light	33.8	150–300	14.2	57.3/31.1/11.4	53.9	0.040
Sirri	30.9	180–360	15.3	59.0/30.0/11.0	52.0	0.078
Soroosh	18.1	200–320	5.1	49.0/42.2/8.8	59.8	0.021
Iraq						
Basrah Light	33.7	150–350	18.8	—/17.0/10.0	37.0	—
Basrah Heavy	24.7	150–350	14.2	—/9.5/21.0	51.5	—
Kirkuk Blend	35.1	150–350	21.0	—/10.5/22.0	54.5	—
Ivory Coast						
Espoir	31.4	300–400	9.4	43.4/48.2/8.4	65.0	0.036
Kuwait						
Kuwait Export	31.4	170–310	11.9	70.1/20.8/9.1	39.0	0.030
Libya						
Amna (High Pour)	36.1	120–250	7.7	69.9/26.5/3.7	33.9	0.010
Amna (High Pour)		250–330	6.9	60.5/33.6/5.9	45.4	0.020

Properties of naphthas in world crude oils—cont'd

Country and Crude Name	Crude Gravity, °API	Approximate Boiling Range, °F	Naphthas in Crude			
			In Crude, Vol %	P/N/A, Vol %	N + 2A, Vol %	Sulfur, Wt %
Libya—cont'd						
Brega	40.4	70–210	12.4	72.5/25.3/2.2	29.7	0.014
Brega		210–300	10.7	53.1/39.4/7.5	54.4	0.020
Bu Attifel	43.6	160–290	9.0	75.4/18.3/5.9	30.1	—
Bu Attifel		290–390	9.0	75.7/17.8/6.1	30.0	0.002
Sarir	38.3	IBP–360	22.9	60.8/35.8/3.4	42.6	14 wt ppm
Zueitina	41.3	113–220	9.3	65.5/30.9/3.6	38.1	—
Zueitina		220–390	21.4	59.2/27.0/10.2	47.4	0.040
Malaysia						
Bekok[2]	49.1	145–330	26.8	57.4/29.3/13.3	55.9	5 wt ppm
Labuan Blend[2]	33.2	145–330	18.8	17.5/74.6/7.9	90.4	20 wt ppm
Miri Light[2]	36.3	145–330	23.1	14.6/68.9/16.5	101.9	17 wt ppm
Pulai[2]	42.5	145–330	27.9	31.8/26.4/31.4	89.2	<5 wt ppm
Tapis[2]	44.3	145–330	23.5	40.3/31.1/28.7	88.5	13 wt ppm
Tembungo	37.4	160–300	17.9	27.0/59.0/14.0	87.0	0.004
Mexico						
Isthmus	32.8	60–400	29.9	64.4/21.0/14.6	50.2	0.054
Maya	22.0	60–400	19.7	60.6/27.0/12.4	51.8	0.213
Nigeria						
Bonny Light[2]	36.7	C_5–300	20.9	42.6/47.6/9.8	—	0.002
Bonny Medium[2]	25.2	C_5–300	6.3	32.8/54.7/12.5	—	0.003
Brass River	40.9	200–300	16.6	45.0/43.5/11.5	66.5	<0.001
Brass River		300–400	11.5	45.8/38.5/15.8	70.1	0.0087

APPENDIX 219

Escravos	36.2		170–310	16.1	41.6/46.5/11.9	70.3	0.013
Pennington	36.6		C$_5$–200	7.4	70.5/25.5/3.5	32.5	0.001
Pennington			200–340	14.7	35.0/52.5/12.0	76.5	0.001
Qua Iboe	35.8		180–380	25.1	34.6/55.3/10.1	75.5	0.026
Norway							3 wt
Ekofisk	43.4		180–300	20.2	53.1/36.8/10.1	57.0	ppm
Statfjord	38.4		100–360	22.1	48.9/37.3/13.8	64.9	0.0037
Oman							
Oman Export	36.3		220–390	17.8	67.6/21.3/10.1	41.5	0.03
Qatar							
Dukhan	41.7		155–265	11.7	67.8/19.7/12.5	44.7	0.02
Dukhan			265–350	10.7	64.3/13.9/21.1	56.1	0.03
Qatar Marine	35.3		155–265	9.5	60.6/23.3/16.1	55.5	0.03
Qatar Marine			265–350	10.1	56.7/18.5/22.9	64.3	0.04
Saudi Arabia							
Arabian Heavy	27.9		70–210	7.9	88.6/10.2/1.2	12.6	0.0059
Arabian Heavy			210–300	6.8	70.8/19.5/9.7	38.9	0.016
Arabian Light	33.4		70–210	8.9	86.8/10.7/2.5	15.7	0.024
Arabian Light			210–300	8.5	68.3/19.0/12.7	44.4	0.027
Arabian Medium	30.8		70–210	8.9	85.3/12.3/2.4	17.1	0.043
Arabian Medium			210–300	7.7	68.5/18.8/12.7	44.2	0.050
Sharjah, U.A.E.							
Mubarek	37.0		220–300	9.1	51.6/30.4/18.0	66.4	0.01
Mubarek			300–360	7.2	48.5/28.8/22.7	74.2	0.01
Syria							
Souedie	24.9		150–350	13.1	—/18.0/10.5	39.0	—
Trinidad							
Galeota Mix	32.8		C$_5$–200	3.7	63.1/30.5/6.4	43.3	0.010
Galeota Mix			200–320	8.7	39.4/44.9/15.7	76.3	0.017
Tunisia							
Ashtart	29.0		60–210	5.7	—/—/10.0	—	—
Ashtart			210–290	6.4	47.2/37.8/15.0	67.8	0.0055
Ashtart			290–360	5.9	46.5/33.8/19.7	73.2	0.0120

Properties of naphthas in world crude oils—cont'd

Naphthas in Crude

Country and Crude Name	Crude Gravity, °API	Approximate Boiling Range, °F	In Crude, Vol %	P/N/A, Vol %	N + 2A, Vol %	Sulfur, Wt %
United Kingdom						
Beatrice, North Sea	38.7	160–210	3.5	63.7/32.0/4.3	40.6	0.0002
Beatrice, North Sea		210–300	7.3	69.4/18.6/12.2	43.0	0.0004
Flotta, North Sea	35.7	150–300	14.9	66.0/27.5/16.0	59.5	0.001
Flotta, North Sea		300–360	5.6	49.5/30.5/19.5	69.5	0.04
Maureen, North Sea	35.8	60–400	27.9	57.3/32.0/10.7	53.4	0.002
Montrose, North Sea	40.1	200–320	15.0	56.2/33.5/10.3	54.1	0.001
Ninian, North Sea	35.6	65–300	19.9	63.0/27.0/10.0	47.0	0.001
Ninian, North Sea		300–400	9.3	44.0/40.0/16.0	72.0	0.006
Tartan, North Sea	41.7	150–300	20.0	56.5/25.5/17.5	60.5	0.01
Tartan, North Sea		300–360	7.1	50.0/30.0/19.5	69.0	0.06
United States						
Alaska, North Slope[2]	26.4	150–180	15.6	39.7/43.3/17.0	77.3	0.013
Kuparuk, Alaska[2]	23.0	150–380	14.5	38.3/47.0/14.7	76.8	0.018
West Sak, Alaska[2]	22.4	150–380	14.4	36.4/48.2/15.4	80.0	0.018
Camrick, Tex PHDL[3]	40.2	100–430	30.4	62.1/30.1/7.8	45.7	0.013
Camrick, Tex PHDL		220–360	11.7	56.4/36.2/7.5	51.2	0.010
Camrick, Tex PHDL		320–430	8.4	58.5/30.0/11.5	53.0	0.03
Brazoria, Gulf Coast[3]	37.6	100–400	28.6	48.0/39.3/21.7	82.7	0.003
Brazoria, Gulf Coast		210–320	11.1	49.4/37.8/12.8	63.4	0.001
Brazoria, Gulf Coast		310–400	10.7	44.4/40.5/15.1	70.7	0.005
Ellenberger, West Texas	47.6	210–420	24.8	72.2/22.0/5.8	33.6	0.08
North Cowden, West Texas Sour[3]	33.7	90–400	32.7	58.9/30.5/10.6	51.7	0.21
North Cowden, West Texas Sour[3]		210–300	13.4	46.0/36.4/17.6	71.6	0.14
North Cowden, West Texas Sour[3]		300–400	9.4	54.7/33.3/12.0	57.3	0.49
Hondo, California[3]	19.3	100–390	18.2	55.2/37.4/7.4	53.2	0.87
Hondo, California		220–310	5.7	48.6/40.2/11.2	62.6	0.52
Hondo, California		310–410	5.9	45.8/44.1/10.2	64.5	1.74
Kern River, California[4]	13.7	220–420	1.0	0.0/92.0/8.0	108.0	0.1

Name						
San Joaquin Valley, California[4]	34.4	220–340	17.0	33.0/57.0/10.0	77.0	0.03
Los Angeles Basin, California[4]	21.7	220–350	9.5	25.0/67.0/8.0	83.0	0.08
Oklahoma Average[4]	39.9	270–400	15.8	50.0/40.0/10.0	60.0	0.014
Louisiana[3]	38.0	120–420	16.3	60.0/27.6/12.4	52.4	0.009
Louisiana		210–340	6.1	55.3/30.5/14.2	58.9	0.003
Louisiana		310–430	6.6	58.2/29.4/12.5	54.4	0.02
Louisiana Delta[4]	30.6	310–375	5.8	34.4/52.7/12.9	78.5	0.1
Rocky Mountain Sweet[4]	39.3	280–400	14.4	46.0/43.0/11.0	65.0	<0.001
Wyoming Sour[4]	24.9	IBP–280	9.4	65.0/28.0/7.0	42.0	0.10
Wyoming Sour		280–400	10.3	51.0/35.0/14.0	63.0	0.13
Wyoming Int. Sour[4]	35.0	260–400	14.1	54.0/33.0/13.0	59.0	0.11
USSR						
Soviet Export Blend	32.5	190–360	12.8	38.5/52.0/9.5	71.0	0.029
Venezuela						
BCF 24	23.5	200–300	5.0	48.2/39.9/11.9	63.7	0.0066
Bachaquero[2]	16.8	200–300	3.5	29.4/59.0/11.7	—	0.026
Bachaquero		300–350	1.8	23.2/59.6/17.2	—	0.1107
Boscan	10.1	200–300	1.3	53.3/—/—	—	0.68
Cueta	31.8	80–200	7.5	75.7/19.9/4.5	28.9	0.002
Cueta		200–300	9.9	50.9/32.1/17.1	66.3	0.01
Lagomedia	31.5	80–200	6.1	77.4/20.4/2.3	25.0	—
Lagomedia		200–300	9.3	55.4/33.0/11.5	56.0	0.0
Lagomedia		300–350	5.0	50.0/32.5/17.5	67.5	—
Leona	24.1	200–300	6.7	49.9/32.1/18.0	68.1	0.005
Merey	17.4	200–300	3.2	43.9/41.6/15.2	72.0	0.01
Merey		300–350	1.9	43.2/42.3/14.5	71.3	0.036
Tia Juana Light	32.1	200–300	10.8	66.0/24.3/9.7	43.7	0.0017
Tia Juana Light		300–350	4.8	51.2/31.7/17.1	65.9	0.0077
Zaire						
Zaire	31.7	170–310	8.8	53.8/41.6/4.6	50.8	0.006

[1]Aalund, Leo. "Guide to Export Crudes for the 80s." *OGJ* April–December 1983. © 1983 PennWell Publishing Co.
[2]P/N/A, weight %
[3]Data from sources other than ref. 1 and 4.
[4]"Evaluation of World's Important Crudes." *OGJ* 1973 Reprint, © 1973 PennWell Publishing Co.

GLOSSARY

ABD Average bulk density. Generally the weight of catalyst per unit volume. For example, pounds per cubic foot, which can be expected when loading a commercial reactor with catalyst. A laboratory method of pouring catalyst into a graduated cylinder, without vibration, is sometimes reported as ABD.
Activity See catalyst activity.
Adiabatic reactor Heat is neither added nor removed in the catalyst bed.
Alumina Aluminum oxide, Al_2O_3. Gamma alumina and eta alumina are used for a base or support for catalytically active materials in reforming catalysts.
Aromatics A classification of hydrocarbons distinguished by a benzene ring or rings in their molecular structure. Aromatics associated with catalytic reforming are benzene, toluene, orthoxylene, metaxylene, and paraxylene.
Barrel U.S. petroleum industry standard of 42 U.S. gallons at 60°F and 14.696 psia.
Benzene A six-carbon atom molecule of ring structure with three double bonds, usually represented as

$$\begin{array}{c} H \\ | \\ C \\ H-C \diagup \diagdown C-H \\ | \quad \| \\ H-C \diagdown \diagup C-H \\ C \\ | \\ H \end{array} \quad \text{or, for simplicity:} \quad \bigcirc$$

Bimetallic See catalyst.
Blending octane number Octane number computed for one component which will make the calculated octane number of the total blend equal to the laboratory-determined octane number of the blend.
Boiling See lower boiling.
BTX reformer A catalytic reformer. The primary objective is to produce benzene, toluene, and xylenes for the petrochemical market.
Catalyst A substance that alters or changes the chemical reaction rate but is not itself one of the final products of the reaction. Reforming catalyst is the material in reforming reactors which promotes the chemical reactions for converting feedstock into desired products. Reforming catalysts are cylindrical extrudates or spheres. Diameter is 1/16 in. or 1/18 in. Catalysts are classed as monometallic, bimetallic, or multimetallic, depending on the number of active metals present.
  Monometallic A catalyst that contains one metal promoter, usually platinum.

Bimetallic A catalyst that contains two metal promoters. The most common is a platinum and rhenium combination. Other combinations, such as platinum plus iridium or platinum plus germanium, are in the U.S. patent art.

Multimetallic A catalyst that contains more than two metal promoters, for example, platinum, iridium, and gold.

Catalyst activity A measure of how well a catalyst performs with respect to reaction rate, temperature, or space velocity. A high-activity catalyst operates at a lower reactor temperature or a higher space velocity than a low-activity catalyst.

Catalyst life In reforming there are two ways of expressing catalyst life. *Cycle life* is the period between catalyst regenerations. *Ultimate life* is the period between the start-up with new catalyst and the discarding of it for metals recovery. In either case, life is expressed in time (hours, days, months, years) or in barrels charged per pound of catalyst.

Catalyst-oriented packing See dense loading.

Catalyst selectivity A measure of a catalyst's ability to yield desired products from reforming. A catalyst with good selectivity yields a high percentage of desired octane number reformate or of desired aromatic.

Catalyst stability A measure of the ability of a catalyst to maintain high activity and good selectivity over a long period of time. A catalyst with good stability has a long cycle life between regenerations.

Catalytic reforming A refinery process designed to a) increase the octane number and b) produce aromatics in the gasoline boiling range. The chemical reactions (except hydrocracking) rearrange molecules without changing the number of carbon atoms in the molecule.

Cat reformer Catalytic reforming unit.

Centerpipe A perforated or slotted pipe in the center of a radial-flow reactor, extending from the top to the bottom of the catalyst bed. Vapor flowing radially through the catalyst from the outer annulus collects in the centerpipe and flows out of the reactor. The pipe is usually covered with wire screen to prevent catalyst migration.

Centerwell A fractionator tray designed to trap liquid for withdrawal from the vessel. The tray is usually about half the distance between the bottom and the top of the vessel and is thus called a "centerwell."

Channeling Uneven distribution of flow through a catalyst bed so that some of the catalyst is not contacted by flowing vapor.

Clear Unleaded, when used in reference to gasoline.

Coke General name for the carbonaceous material deposited on the catalyst. Coke can be carbon or a complex mixture of ring-type molecules sometimes called polynuclear aromatics.

Continuous catalyst regeneration A reforming process designed to continuously or to intermittently move the catalyst through the reactors. The catalyst is withdrawn from the last reactor, regenerated in a regeneration section, and returned to the reaction section. Catalyst can be withdrawn or added as desired.

Cracked naphtha Naphtha from refinery units in which hydrocarbon molecules were cracked or split, for example, thermal cracking, catalytic cracking, coking, and visbreaking. Cracked naphthas usually contain a substantial quantity of monoolefins and diolefins.

Crude distillation Crude oil received at a refinery is a mixture of hydrocarbons with boiling points ranging from below 100°F to above 1,100°F. A refiner, by distillation, separates the hydrocarbons into individual streams with boiling ranges suitable for further processing. Typical crude fractions are as follows:

	Boiling Range, °F
Light st.-run gasoline	80–180
Heavy st.-run gasoline	180–400
Distillates (jet fuel, kerosene)	400–500
Diesel	400–650
Atmospheric gas oil	400–650
Atmospheric residuum	650+
Vacuum gas oil	650–1,100
Vacuum residuum	1,100+

The initial boiling points and end points vary, depending on refinery configuration. The volume of each fraction varies, depending on type of crude oil.

Crush strength The weight required to crush a catalyst particle. A laboratory method determines the crush strength, in pounds, on individual catalyst particles and reports the average of a number of determinations.

Crystallite size The average diameter, in angstroms, of metals dispersed on the catalyst. Usually determined by X-ray diffraction.

Cycle life The period between catalyst regenerations. Expressed in time (hours, weeks, or months) or in barrels per pound of catalyst.

Cyclic reformer A reformer with an extra reactor so that the catalyst is regenerated in one reactor while the others continue reforming. Cycles range from a few hours to more than a month.

Cyclization or dehydrocyclization The formation of a ring molecule from a straight-chain molecule.

Cycloversion A naphtha desulfurization process licensed by Phillips with very mild reforming.

Deactivation Loss of catalyst performance when compared to the catalyst's performance in new or fresh condition. The loss can be in yield, octane number, or reactor temperature.

Dehydrogenation Removal of hydrogen from a hydrocarbon molecule.

Delta The difference between the temperature into and the temperature out of each reactor. The change in temperature across a reactor, that is, the difference between the temperatures at the inlet and outlet of each reactor. Total delta T is the sum of the individual reactor delta T's.

Dense loading Also called catalyst-oriented packing (COP). A method of loading catalyst in reactors that results in more pounds of catalyst per volume than are obtained by pouring or by "sock"-loading the catalyst.

Distillation See crude distillation.

Draeger tubes Calibrated glass tubes containing chemicals for quick analysis for specific components of gas streams. For example, hydrogen sulfide, hydrogen chloride, and ammonia.

Dual function Two functions built into a single catalyst. In reforming, a catalyst which promotes chemical reactions attributed to a metal (dehydrogenation) and chemical reactions attributed to an acid (isomerization).

Endothermic A reaction which absorbs heat. In reforming, an endothermic reaction usually results in the vapor leaving the reactor at a lower temperature than the vapor entering the reactor.

Exothermic A reaction which releases heat. In reforming, an exothermic reaction can result in a temperature rise from vapor entering to that leaving a reactor.

Ex-situ regeneration See merchant regeneration.

Extract One of the products of the extraction unit. See raffinate.

Fractionation A distillation process carried out in a vessel such that distilled vapors rising up the vessel are contacted with liquid flowing down the vessel. The contact takes place on trays at intervals within the vessel. Means are provided to draw liquid off certain trays. Each liquid portion drawn off is a fraction of the material being distilled.

Fractionator A vessel designed for fractionation.

Fresh catalyst A reforming catalyst that has just been regenerated is "fresh"—to differentiate it from a catalyst which has never been used.

Hydrodenitrification Removal of nitrogen from a hydrocarbon molecule in the presence of hydrogen.

Hydrodesulfurization Removal of sulfur from a hydrocarbon molecule in the presence of hydrogen.

Hydrogen make Hydrogen yield from the reforming unit. Expressed in wt % of feed or in standard cubic feet per barrel of feed (scf/bbl).

Hydrogen purity Hydrogen content of recycle gas or of hydrogen make, usually in mol percent.

Hydrogen recycle Sometimes a synonym for total recycle gas. Other times hydrogen recycle means only the hydrogen content of the recycle gas.

Hydrogen-to-hydrocarbon ratio, H:HC Ratio of mols of hydrogen to mols of naphtha charge. Hydrocarbons in the recycle gas are not included in the ratio.

Hydrotreating A general term for refinery processes in which the hydrocarbon feed contacts the catalyst in the presence of hydrogen but at conditions which usually do not cause cracking or splitting of molecules into lower boiling hydrocarbons. For example, hydrodesulfurization removes sulfur from a molecule to form hydrogen sulfide, replacing the sulfur in the molecule with hydrogen. The same applies to hydrodenitrification. If the sulfur or nitrogen is part of a ring molecule, the ring, which must be broken to remove the sulfur or nitrogen, may not undergo further splitting but forms a paraffin.

IFP reforming A reforming technology developed and licensed by IFP.

In situ regeneration In place regeneration. Regeneration of catalyst without removing it from the reactors.

Isomerization Rearranging the structure of a molecule without altering its molecular formula. In a petroleum refinery, isomerization is achieved with aid of a catalyst and sometimes in the presence of hydrogen.
Lighter streams See lower boiling.
Loading density The density, in pounds per cubic foot, calculated after loading a commercial reactor. See ABD.
Lower boiling A refinery stream having a lower initial boiling point than the stream to which it is compared.
LPG Liquefied petroleum gas. Usually propane, butane, isobutane, or a mixture of these. Generally used as fuel for heating.
Magnaforming Reforming technology licensed by Engelhard.
Makeup Replacement for catalyst fines removed when the catalyst is screened.
Merchant regeneration Ex-situ regeneration. Regeneration of a catalyst that has been removed from the reformer reactors and sent to a catalyst regeneration service. Only carbon burnoff at a controlled temperature is done. The catalyst is reactivated by the refiner after it is reloaded into the reactors.
Methyltertiarybutylether $C_5H_{12}O$. A high-octane-number, 110 $(R + M)/2$, blend stock produced from methyl alcohol and isobutylene.
Monometallic See catalyst.
Motor fuel pool The total of all gasoline produced in a refinery for sale as gasoline.
Motor fuel reformer A catalytic reformer operated to raise the octane number of the naphtha feed.
Multimetallic See catalyst.
N + 2A Used for characterization of reforming feedstock. It is the naphthene plus two times the aromatic content in volume percent.
Naphtha A refinery stock within the boiling range of 80–435°F. Naphtha may be in a narrow-boiling-point range, 180–220°F, or in a wide-boiling-point range, 180–400°F. Naphtha can be straight-run or virgin (not subjected to thermal or catalytic cracking), or it can be cracked. Naphthas are generally gasoline-boiling-range stocks which are not blended directly to gasoline, but which are charged to a processing unit such as a cat reformer or a hydrotreater.
Nonregenerative A reformer which does not regenerate catalyst in place but which replaces spent catalyst with new or with merchant-regenerated catalyst.
NPRA National Petroleum Refiners Association, Suite 1000, 1899 L Street NW, Washington, DC 20036.
NPRA Q&A NPRA Question-and-Answer sessions on refining technology.
Octane number The measure of antiknock performance of gasoline. Refiners use two octane-number methods. The research method (reported as RON), ASTM D–2699, simulates engine performance at low speeds in a laboratory. The motor method (reported as MON), ASTM D–2700, simulates engine performance at high speeds. The octane number on most pumps at service stations is the arithmetic average of the two methods, $(R + M)/2$.
On stream samplers Devices installed in reforming reactors which permit withdrawal of a small quantity of catalyst without shutting down the unit.
Platforming Reforming technology licensed by UOP.

Poisons Substances that undesirably alter the performance of a catalyst. Two types of poisons in reforming are defined below.
> *Temporary poisons* cause a catalyst to behave abnormally. When these poisons are eliminated from the feed, the catalyst recovers most, if not all, of its desired characteristics. Sometimes regeneration of the catalyst is necessary. Examples of temporary poisons are sulfur and nitrogen.
>
> *Permanent poisons* cause an irreversible loss in catalyst performance. The catalyst cannot be restored to usable condition and must be removed from service. Heavy metals such as arsenic and lead are permanent poisons.

Pool octane See total pool octane.

Powerforming Reforming technology licensed by Exxon.

Precursor A molecule which is most likely to be converted to a molecule of different configuration in reforming. For example, cyclohexane is a precursor of benzene because, at reforming conditions, cyclohexane is converted nearly 100 mol % to benzene.

Product separator A vessel for flash separation of effluent from the last reactor. The flashed vapor is mostly hydrogen, methane, and ethane; the liquid is mostly pentane and heavier hydrocarbons.

Proof burn Part of the catalyst regeneration procedure to remove the last traces of carbon from the catalyst.

(R + M)/2 Arithmetic average of research and motor octane numbers.

Radial flow reactor A reactor in which vapor flows from a peripheral annulus through the catalyst bed to the center of the reactor into a centerpipe, through which it flows out of the reactor.

Raffinate The paraffin-rich product remaining after extraction of aromatics from the reformate.

Raw naphtha Naphtha charge to a hydrotreater. Untreated naphtha.

Reactivation Sometimes rejuvenation. After regeneration, the redispersion of precious metals on the catalyst by the catalyst's contact with chloride and oxygen. This treatment restores catalyst activity to that of fresh catalyst.

Reactors Vessels containing the catalyst.

Recycle gas Hydrogen-rich gas flashed off the product separator and circulated to the reactor system by the recycle compressor.

Reformate The pentane-and-heavier hydrocarbon yield from a catalytic reformer, usually identified as C_5^+ reformate.

Regeneration Burning carbon off the catalyst to restore its activity. Regeneration sometimes means the entire procedure, including carbon burn, redispersion of metals, reduction, chloriding, and sulfiding.

Reid vapor pressure Vapor pressure at 100°F determined by ASTM method D–323.

Rejuvenation See reactivation.

Rheniforming Reforming technology licensed by Chevron.

Scallops An arrangement of slotted plates which holds the catalyst away from the reactor shell to form the annulus for vapor flow. When installed in a reactor and viewed from above, these plates resemble scallops.

Screening Passing of the catalyst through a series of selected wire mesh screens which sort catalyst particles, fines, and extraneous material by size.
Selectivity See catalyst selectivity.
Semiregenerative A reformer that operates an extended period between catalyst regenerations. The period (cycle) may be from six months to more than three years. The catalyst normally is regenerated in situ in the reactors.
Sintering of the catalyst is growth or agglomeration of metal crystallites.
Slop A catch-all term for refinery hydrocarbon streams collected and accumulated from leaks, emergency shutdowns, equipment drainage for maintenance, or just being off specification.
Slump-and-seal catalyst Catalyst placed on top of the catalyst bed to allow for settling of catalyst during reforming operation.
Stability See catalyst stability.
STAR Steam Active Reforming. Reforming technology licensed by Phillips.
Stoichiometric Quantitative yield derived from the equation of a chemical reaction.
Stream A flowing fluid, either liquid or gas.
Sulfiding Sometimes called presulfiding. Passing a sulfur compound such as ethyl mercaptan or dimethyl sulfide over the catalyst in the presence of hydrogen. The sulfur combines with the catalyst, especially on the superactive sites, and helps control reactions on start-up of a unit with fresh or regenerated catalyst.
Support material Inert material (or sometimes catalyst) in the forms of balls, pellets, granules, or broken bricks placed below a catalyst bed as fill in the bottom of a reactor. Support material is also placed on top of a bed of catalyst to distribute vapor flow and protect the catalyst from the impact of high-velocity vapor.
Swing reactor A reactor which can be isolated from the rest of the equipment. It permits either in situ regeneration of the catalyst or performance of maintenance without a shutdown of the rest of the unit.
Tetraethyl lead Used as an octane-number booster in gasoline.
Toluene An aromatic hydrocarbon consisting of a benzene ring with one methyl group:

Total pool octane Average octane number of all gasoline produced for sale as gasoline.
Trash baskets Slotted or perforated baskets (pipe) imbedded in the top of the

catalyst bed to collect scale and fines. The purpose of these baskets is to prevent blockage or plugging of the top layer of catalyst.

Troy ounce The system used for reporting the weight of precious metals such as platinum and palladium. Reforming technologists use a factor of 14.583 troy ounces per pound avoirdupois.

Ultimate life The life of catalyst from initial use as new catalyst to removal for metals recovery on discarding. Expressed in time (hours, days, months, years) or in barrels per pound.

Ultraforming Reforming technology licensed by Amoco.

Unleaded See clear.

Weighted average bed temperature Calculated from the average of the inlet and outlet temperatures of each reactor and the weight percent of the total catalyst in each reactor.

Weighted average inlet temperature Calculated from the inlet temperatures of each reactor and the weight percent of the total catalyst in each reactor.

Xylenes Benzene rings with two methyl groups:

orthoxylene   metaxylene   paraxylene

Yield deviation The difference in volume percent of C_5^+ reformate yield being obtained and that which would be expected with fresh catalyst, adjusting for feed properties and operating conditions.

Index

A

Alkylate blending octane number 11
Alumina 60, 61
  base for catalyst 61
  eta 61
  gamma 61
  transformation 61
Aromatics 1, 2, 12, 26, 142, 143, 147, 148, 149, 150
  analysis in reformate 150, 151
  benzene 12
  blending octane number 12
  distribution in reformate 142, 148
  n-heptane–toluene equilibrium 144
  separation from reformate 1
  sources 142
  toluene 12
  uses 142
  xylene 12
  yields from BTX reforming 144, 243
Arsenic (see catalyst poisons)
  adsorption by HDS catalyst 115

B

BTX reforming 142–152
  feedstock composition 147, 148, 149
  feedstock paraffins, effect of 144, 145
  feedstock selective 143
  measure of performance 148, 149
  material balance 147
  purpose 1
  temperature, effect of 143, 144
  yields 26, 143, 144
  variables 145, 146
Butane and lighter
  isobutane from n-butane in feed 98
  yields, estimating 99, 100

C

Catalyst 40–72
  acid function reactions 59, 60
  activity 41, 52

Catalyst—cont'd
  base material 57, 58
  bifunctional 59
  bimetallic xvi
  channeling of vapor through 38
  chloride promoter 57, 60
  coke buildup 108
  commercial 41, 42, 61, 62
  composition 60, 61
  COP, catalyst-oriented packing 67
  crush strength 68
  crystallites 64
  cycle life 56, 57
  density, ABD 64
  density, loading 64
  dense loading, COP 67
  development 41, 57, 58
  different types in one system 91
  dual function 59
  fluoride promoter 57, 60
  manufacturers 209
  migration out of reactor 111
  metals content 62, 63
  metals function 59, 60
  metals recovery from 65, 66
  multifunctional 59
  patents 62, 63
  platinum on alumina 58
  platinum cost 64, 65
  platinum requirement 64
  plugging of catalyst bed 38
  poisons (see catalyst poisons)
  preparation 58
  pressure drop through 67
  properties 61
  rejuvenation (see regeneration)
  rhenium cost 65
  selectivity 54
  samplers, on stream 106, 107
  sintering 112
  shapes 57, 67
  sizes 57, 67

232 INDEX

Catalyst—cont'd
  silica-alumina base 60
  stability 55, 56
  state of art, 1980s 59
  sulfur content in new 62
  sulfiding 62
  support, top and bottom of bed 36
  surface area 66
  surface area loss 66
  ultimate life 33
  vapor bypassing 36
  volume to load reactor 64
Catalyst poisons 68–70, 110–113
  arsenic 68, 69, 70, 113
  coke 68, 70, 111
  copper 69, 70, 113
  effect on delta T 113
  halides 69
  lead 68, 69, 70, 113
  nitrogen 68, 69, 110
  permanent 68, 113
  phosphorous 69
  silica 69
  sulfur 68, 69, 110
  temporary 68, 69, 110
  water 70
Catforming, first unit xv
Centerpipe 35, 36
Channeling (see catalyst)
Chloride
  effect on activity of catalyst 106
  control 106
  determination by Draeger tube 108
  equilibrium on catalyst 106
  hydrocracking, caused by 108
  injection 31
  overchloriding 108
  effect on selectivity 106
  underchloriding 109, 110
  water and chloride 106
Chronology of reforming xv
Classification of reforming 33, 34
  continuous catalyst regeneration 34
  cyclic 33
  moving bed 34
  nonregenerative 33
  semiregenerative 33
Coke 76
  feedstock endpoint effect 76
  on catalyst after carbon burn (see regeneration)
Commercial processes 153–173
  IFP catalytic reforming 154–158
  Magnaforming 158–160
  Platforming 160–163
  Powerforming 163–166
  Rheniforming 166–168

Commercial processes—cont'd
  Ultraforming 169–171
  STAR 171–174
Compressor 38, 111, 116
Computer modeling of reforming 19
Continuous catalyst regeneration 34, 160, 163
  catalyst circulation 34
  features 34, 160, 163
  installed U.S. capacity 34
  reactors 34
Corrosion 116
Cracked naphtha 78, 116–123
  cat cracked 118
  coker 118
  feedstock for reforming 78, 116, 118, 119
  hydrogen consumed in hydrotreating 122
  hydrotreating 122, 123
  N + 2A 118
  nitrogen content 123
  octane loss from hydrotreating 122, 123
  octane number 118, 119
  olefin content 78, 117
  pyrolysis naphtha 118
  reforming 123–127
  sulfur content 123
Cycle life 33, 106
Cycloversion xiv
Cyclic reformer 33, 34, 35

D

Dehydrocyclization 17, 18
Dehydrogenation 15
Delta T 112–114
  arsenic, effect on 114
  catalyst poisons, effect on 113
  hydrogen purity, effect on 113
  H:HC ratio, effect on 113
  lead, effect on 114
  loss of, causes 112
Demethylation reaction 22

E

Exchangers 31, 110
  energy conservation 31
  leak detection 110
Engelhard (see Magnaforming)

F

Feedstocks 31, 73–75
  boiling point of hydrocarbons 77
  boiling point 76
  characterization for reforming 24
  composition 22, 23, 24
  cracked naphthas (see cracked naphtha)
  end point 76

Feedstocks—cont'd
  initial boiling point 76
  N + 2A, calculated 75
  PNA distribution 25
  properties, effect on yields 74, 75
  reactor charge 31
  total reactor charge 31
Feed stripper 29, 30
Flow scheme for reformer 28, 29
Fouling of equipment 116
  ammonium chloride 116
  ammonium sulfide 116
  iron sulfide 116
  oxygen 116
  polymer 116, 117
  prevention 116
  water wash to remove deposits 117

G

Gasoline
  distribution by grades 6, 7
  distribution, future 6, 7
  lead content limit 7
  octane numbers 6, 7
  refinery pool blend 9
  refinery pool octane number 9
  unleaded 4

H

Heaters, reformer 32
n-Heptane
  isomerization 18
  hydrocracking 18
Hexanes as reformer feed 3
Houdriforming xv
Houdry catalytic cracking xiii
Houdry, Eugene xiii
Hydrocarbons in straight-run naphtha 210-213
Hydrocrackate
  as reforming feedstock 128
  octane number 128
Hydrocracking 18, 19, 108, 109
  butane produced 19
  chloride promotes 108
  indications of 108
  of n-heptane 18
  pressure effect 19
  propane produced 19
  water injection causes 109
  temperature effect
Hydrodesulfurization of naphtha 175-194
  arsenic on catalyst 184
  ammonium chloride deposits 189
  catalyst activity 182, 183
  catalyst coking 182
  catalyst composition 181, 182

Hydrodesulfurization of naphtha—cont'd
  catalyst deactivation 182, 183
  catalyst density 182
  catalyst life 183
  catalyst shape 181
  catalyst size 181
  catalyst sulfiding 185
  catalyst support 191-193
  catalyst support, guidelines 193
  chemical reactions 176-178
  corrosion of equipment 189
  deposits on catalyst 189
  distillates in feed 175
  of disulfides 176, 177
  equipment failure 190
  fouling of equipment 181
  fouling top of catalyst bed 181
  hydrogen consumption 176-178
  lead on catalyst 184
  mercaptans in feed 176, 177
  monitoring unit performance 186
  operating conditions 180
  olefins 178
  organic chlorides in feed 190
  Phenol 178
  pressure differential of reactor 181
  process design 176, 179
  pyridine 178
  reactor design 181
  recombination of hydrogen sulfide and olefins 187
  reaction rates 176
  regeneration of catalyst 182, 183
  safety 190, 191, 192
  silica deposits 188
  skimming catalyst bed 181
  sulfiding catalyst 185
  sulfur removal efficiency 180
  temperature rise 176
  trash baskets in top of catalyst bed 181
  troubleshooting 186-188
Hydroforming xiii
Hydrogen
  consumption in HDS unit 177, 178
  flammability in air 137
  recycle 31
  steam methane as source 1
  uses in a refinery 1
  yield from reforming 1, 99
Hydrogen-to-hydrocarbon ratio, H:HC 90, 91, 113
  benefits of reduction 91
  calculation 90
  carbon on catalyst, effect 90, 91
  delta T, relation to 113
  division 91, 158
  prevents coking 91

234 INDEX

Hydrogen sulfide, determination 111
Hyperforming xv

I

IFP reforming (see commercial processes)
Isomerization 16, 17, 57
  halogens promote 57
  of naphthenes 17
  of paraffins 16
  reaction 16

K

Kinetics of reforming reactions 19

L

Licensors of reforming technology 153–174, 208
  Amoco (Ultraforming) 169
  Chevron (Rheniforming) 166
  Engelhard (Magnaforming) 158
  Exxon (Powerforming) 163
  IFP (IFP reforming) 155
  Phillips (STAR) 171
  UOP (Platforming) 160

M

Magnaforming xv, 158, 208
Material balance 92–97
Merchant regeneration of catalyst 136, 183
Methycyclohexane 15
Methycyclopentane 17, 55

N

N + 2A, yield-octane correlation 24, 25
  calculated 75
Naphtha
  cracked naphthas 3
  desulfurization 3
  heavy straight-run 3, 210–214
  light straight-run 3, 210–214
  in world crude oils, properties of 215–221
Naphthene conversion 17
Nitrogen as catalyst poison 110
Nonregenerative reforming 33, 35

O

Octane number 4, 7, 10, 11, 20, 21, 110, 111
  blending octane number 11
  motor method 4
  loss of 11, 110
  (R + M)/2 4
  ratings on service station pumps 4
  refinery pool 7, 10
  reforming reactions, effect on 20, 21
  research method 4

Octane number—cont'd
  of selected hydrocarbons 4
  significance 4
  standards for rating 4
Olefins, octane number of 79
  in reformate 19
Orthoforming xv

P

Pentanes as reformer feed 3
Petrochemicals, physical properties 198–201
Platinum 64, 65, 136
  cost 64, 65
  retained on catalyst during regeneration 136
Platforming xv, 20, 21, 145, 154, 160-163
Powerforming xv, 53, 163
Pressure 87, 88, 89, 90, 111
  average reactor 87
  affects coking of catalyst 89, 90
  affects cycle life 89
  affects hydrogen yield 88
  affects reformate yield 88
  affects reforming reactions 87
Pressure differential across reactor 88, 111
Process variables 73, 98–106
  plotting of data 98–106
Product separator hydrogen content 32
  liquid and gas composition 32
Poisons (see catalyst poisons)
Pyrolysis naphtha 118

R

Reactions of reforming 15–27
  process variables, effect on 19
  relative rates 19
  yield-octane from 20, 21
Reactors 29–38, 136–138
  arrangement 29, 31
  cold wall 37
  deflector plate 35
  downflow 37, 38
  hotwall 37
  internal design 36
  internals, inspection intervals 136
  inlet distributor 37, 38
  isothermal 52
  number of 29, 31, 32
  pilot unit 52
  plugging of catalyst bed 111
  pressure differential 38
  radial flow 31
  revamping 31
  scallops 36
  screens 36
  shell temperature 37

INDEX

Reactors—cont'd
  skin thermocouples 138
  spherical 38
  trash baskets 37, 38
  upflow 38
Recombination of sulfur and hydrogen sulfide 187
Refinery process flow scheme 2, 13
  with motor fuel reforming 2
  with BTX reforming 13
Reformate 77, 97, 115
  boiling range 77
  color 115
  yield deviation 97
Regeneration of catalyst 33, 66, 129–136
  carbon burn 129, 131, 134, 136
  chemicals required 135
  chloride injection 131
  coke on catalyst 134
  connections for 33
  corrosion protection 131
  delay for maintenance 133
  at high pressure 133
  liquid oxygen replacing air 133
  merchant regeneration 136
  metals retention on catalyst 136
  oxygen control 130
  parallel reactor burn 131
  poor flow distribution 133
  proof burn 133
  reduction of catalyst 134
  reduction of hydrogen purity 134
  rejuvenation of catalyst 134
  sequence of steps 129
  sulfiding catalyst 134
  surface area loss 66
  temperature rise 130
  temperature runaway 130, 131
  unloading catalyst 129
  utilities required 135
  water in recycle gas 135
Rheniforming xv, 145, 149, 166–168

S

Safety 137, 138
  hydrogen flammability in air 137
  reactor shell, high temperature 37, 138
  temperature runaway 138
Scallops, reactor 36
Screens, reactor 35, 36
Semiregenerative reforming 33, 35
Simulation (computer modeling) of reforming 19
Sovaforming xv
Space velocity 52, 83–87, 111
  commercial unit range 87
  interchange with temperature 84, 85

Space velocity—cont'd
  liquid hourly, LHSV 52, 83
  minimum 87
  relation to maximum reactor charge 86
  relation to residence time 83
  effect on severity of reforming 84
  temperature runaway, caused by 111
  weight hourly 52, 83
STAR (steam active reforming) 171–173
Sulfiding of catalyst 62, 134

T

Temperature 25, 31, 32–35, 52, 79–83, 97, 106, 203–206
  average reactor bed (WABT) 25, 52
  average reactor inlet (WAIT) 25, 52, 79, 80, 81
  control 31
  conversion tables 203–206
  delta T 82, 83, 97, 106
  end of cycle 33
  endotherms (reactor) 32
  exotherms (reactor) 130, 131, 138
  interchange with space velocity 83
  profile (reactor) 82
Tetraethyl lead 114, 115, 184 (see catalyst poisons)
  adsorption by catalyst 115
  distribution in crude unit fractions 114
  removal from naphtha 184
  thermal stability 114
Trash baskets (reactor) 37, 38
Thermofor catalytic reforming xv
Troubleshooting 110, 111
  bypassing catalyst 111
  octane number loss 110
  yield loss 112
Toluene xiv, 15, 142, 147, 148

U

Ultraforming xv, 169, 170, 208
UOP xiv, 20, 21, 25, 145, 160–163, 208

W

Water 109
Wire mesh sieve openings 202

X

Xylenes 1, 142, 147, 148, 151

Y

Yield (reformate) 12, 15–27, 54, 88, 112, 124, 127
  correlations 20, 21, 25, 54
  cracked naphtha reforming 124, 127
  loss of 112
  pressure effect 88

Donald M. Little is a refinery catalytic process engineering consultant. He received a B.S. in chemical engineering from the University of Nebraska and is a registered professional engineer in Oklahoma. The author has 42 years' refining experience with Phillips Petroleum Co.

At his retirement in 1984, Don Little was Catalytic Processes Principal Engineer, Refining Div. His service with Phillips covered various phases of refinery processing: process design, economics, planning, and engineering for catalytic reforming; crude distillation, high-vacuum distillation, hydrofluoric-acid alkylation, hydrotreating, selective hydrogenation, hydrogen processing, isomerization, catalytic polymerization, and visbreaking. At one time he monitored the operation of 11 catalytic reformers in seven Phillips plants.

Little is coauthor of technical articles on thermal cracking, high-vacuum processing, and foaming of residuum. He served sixteen years on the NPRA Screening Committee for the NPRA Q & A Session of Refining and was a panel member in 1974.

For over 17 years, Little was Refining Div. patent coordinator for Phillips. He is inventor or coinventor of 17 U.S. patents on refinery processes.